Y0-BSQ-654

The Chinese Painted Quail
"The Button Quail"
Their Breeding and Care

**All You Will Ever Need To Know
About This Popular Little Quail**

By

LELAND B. HAYES, Ph.D.

THE CHINESE PAINTED QUAIL

(The Button Quail)

Their Breeding and Care

Copyright © 1992 by Leland B. Hayes, Ph.D.

All Rights Reserved.

No portion of this book may be reproduced in any way without the written permission of the publisher, except for brief excerpts in reviews, etc.

Printed in the United States of America. This book is available to dealers, retailers, libraries and bird clubs in wholesale quantities.

Leland B. Hayes, Ph.D.
P.O. Box 1682
Valley Center, CA 92082
(619) 749-6829

Preface

The Chinese Painted Quail (The Button Quail) - Their Breeding and Care, will be a welcome addition to libraries of the thousands of aviculturists that currently keep and enjoy the marvelous Chinese Painted Quail. This little bird is unique in many ways and is very popular among aviculturists that keep parrot-type birds as they fit well into most any environment. Until this book, there has been no combined source of information on the propagation of the Chinese Painted Quail. This book will be invaluable to the beginning breeder as well as to the more experienced breeder.

Dr. Leland B. Hayes is well qualified to write this book. His latest addition to the literature of upland game birds is a complete work entitled *Raising Game Birds* which along with his wife, Melba, was published in 1987. Other works include, *Basic Quail Propagation, A Basic Guide to Health and Common Diseases of Game Birds*, and many articles appearing in the *Game Bird Breeders Gazette.*

During more than 40 years of aviculture experience, many awards have been given to Dr. Hayes. He has received the coveted Aviculture Hall of Fame Award, the Dr. Gene Delacour Award. He is currently a Director of America's most prestigious game bird organization, the American Game Breeders Cooperative Federation and was recently chosen to be President by the Board of Directors.

This book is a complete book on the subject. It is well organized for easy reading and later reference. The pages are filled with many tips and solid information the reader needs to know. One of the unique things about this book is the dozens of "Hints" that are found throughout each chapter. The four color plates and the black and white photographs will give the reader a good idea of the appearance of the several mutations developed in the Chinese Painted Quail.

We believe that everyone who keeps Chinese Painted Quail should have this book. The information will give not only the needed information to be successful, but will greatly enhance the enjoyment of keeping Chinese Painted Quail. This book is a must for those who want to get into the enjoyment of Chinese Painted Quail.

We highly endorse and recommend this book.

George A. Allen, Jr.
George A. Allen, III.

Editors and Publishers of
Game Bird Breeders Gazette,
Salt Lake City, Utah

Contents

To
These
Friends

There have been so many who have made a contribution to this book.

I must recognize those who gave encouragement and assured me there was a need in the avicultural community for a specialized book just on Chinese Painted Quail.

As always, my endless gratitude goes to George Allen Jr. and George Allen III, for their knowledgeable guidance, and for their priceless friendship through these many years.

Of course behind any success that I have enjoyed is my wife, Melba, who has learned along with me these 33 years of marriage. Without her support and enthusiasm, birds could not have played such an important part in my life.

Leland B. Hayes, Ph.D.
January 9, 1992

Foreword

In the very rewarding hobby of keeping game birds, especially quail, there is developing a firm preference toward the smaller species. This is true because of three distinct reasons:

❑ **Land is very expensive** and fewer people have acreage on which to raise their birds. Most of us live in urban areas and have only small areas for our hobby.

❑ **More cities and townships are adopting ordinances** that prohibit the raising of birds within city limits. This is certainly true in the more populated areas.

❑ **Many people are getting into the wonderful hobby of raising birds**. Some do not want to make a heavy financial investment.

Chinese Painted Quail meet these conditions!

❑ **They require little room.** They are happy in a well vented and lighted garage or in a small backyard.

❑ **They are noiseless.** Neighbors will probably never know that they are present unless the proud owner brags about the hours of enjoyment that they provide.

❑ **They do not require exotic diets.** Their feed is available in any community and is not expensive.

❑ **They do not eat much.** It will take only a few pounds of feed to sustain these little quail for a whole year.

❑ **They are inexpensive.** Any man, woman, or young person that wants to have quail can have them as they will easily fit into most budgets.

❑ **This little book is for you**. The novice or the mature aviculturist will benefit from this book. I have put my basic ideas and experience into plain language. My yearning is for you to decide to join me in keeping the congenial little Chinese Painted Quail and that you will become a triumphant Chinese Painted Quail breeder.

Leland B. Hayes, Ph.D.
Valley Center, California

List of Photographs and Illustrations

List of Advertisers

To be as practical as possible, we have included at the back of this book, advertisements of suppliers that manufacture and sell some of the products that I use and recommend. This information is especially valuable for the beginner.

HINT: Scattered through the book you will find many boxes containing special information which will give emphasis to the things that I consider important for you to remember. These stand out in the book and hopefully they will do the same in your memory as they are designed to help you toward success.

These special thoughts have come from 40 years of avicultural experience. When applied to your situation, hopefully they may indeed help you avoid many of the problems that I have encountered.

The reader is encouraged to make notes as he reads the book. Nothing to me is more aggravating than to read something and then forget where I read it. Some of the most valuable books in my possession are those that have "reminders" for quick reference.

There is plenty of white space for your note taking and several pages have been inserted just for this purpose.

If you like this book, please tell us. Tell your friends about this book. Happy reading!

PLATE A: *A young male silver red breasted mutation.*

PLATE B: *A normal male offering a mealworm to a female.*

15

PLATE C: *A normal male Chinese Painted Quail.*

PLATE D: *A normal male (foreground) and a red breasted male (background).*

16

PLATE E: *A female silver mutation. Notice the barring color pattern.*

PLATE F: *A silver red breasted male (center) and a male red breasted mutation.*

17

PLATE G: *A male red breasted mutation coming out of the moult.*

PLATE H: *Chinese Painted Quail like to dust in sand.*

18

This is a pair of Chinese Painted Quail eating out of a typical feeding pan.

Female Chinese Painted Quail get larger and plumper than males.

19

Group of mutations including White, Silver, and Red Breasted.

Male normal in typical posture.

CHAPTER 1

WHAT IS A
CHINESE PAINTED
QUAIL?

"What a cute little bird," exclaimed the lady in the pet store. "Can it live in the cage with my Cockatiels?" "What does it eat?" The questions to the pet store clerk go on and on. An isolated situation? No, not at all, it happens all the time. Also, more aviculturists are including this little bird in their collections.

This smart looking little quail which caught the eye of the lady in the pet shop is known as the **Button Quail** in America, or the **Chinese Painted Quail** in England and Europe. It's scientific name is *Coturnix chinensis (L.)*. Many game bird breeders in the United States and Europe are now calling this quail the **Chinese Painted Quail** or the **Painted Quail**. It is also known in Asia and some parts of Europe as the

Asian Blue-breasted Quail. It is more often now advertised by the common name **Button Quail** in the US. However, most aviculturists know the true Button Quail to be in the *Hemipode* family and not a true quail at all. Other vernacular names include: **Blue Quail; Blue-breasted Quail; King Quail** in Australia.

I suppose that a **Chinese Painted Quail** by any name is still the same wonderful little bird! We will call this little bird a **Chinese Painted Quail** in this book so our European readers will know which quail we are talking about.

The Coturnix Quail Family

The Coturnix family consists of several species of quail found in Europe, Asia and Africa. Some aviculturists call them *"Old World Quail."* One species lives on the continent of Australia. Their main characteristic is their habit of migration. They go many miles in search of food or to escape the hazards of bad weather. This is the family of quail (Pharaoh or Coturnix) that is so well known in the *Old Testament* which fed the Children of Israel in the wilderness *(Exodus 16:13)*.

This quail is very popular in Europe and Asia as a meat and egg bird. It matures at six weeks of age and will lay an egg everyday of its life if it receives a proper diet. The eggs are pickled and served in bars around the world as a delicacy. The meat is a valuable source of protein in the diets of some of the people in the third world.

The colorful Chinese Painted Quail is a distinguished member of this interesting family. It is not large enough to be eaten so it will never be used on a commercial basis as is the Coturnix (Pharaoh) Quail.

The Coturnix (Pharaoh) Quail has become one of the most popular commercial quail. It is used for meat and egg production.

Distribution map of Asian (AS) and the African (AF) Painted Quail. (Johnsgard)

This photo shows the dusting box and the way the young hens drop their eggs.

Johnsgard lists 10 subspecies of this quail which range from India through the islands to the coast of Australia (Johnsgard, 1988):

- ❑ *Coturnix chinensis chinensis:* India to Malaya, Indochina, and south-eastern China.
- ❑ *C. c. trinkutensis:* Nicobar Islands.
- ❑ *C. c. palmeri:* Sumatra and Java.
- ❑ *C. c. lineata:* Philippines, Borneo, and Celebes.

- *C. c. lineatula:* Lombok, Sumba, Flores, Timor Islands.
- *C. c. lepida:* Islands of Bismarck Archipelago.
- *C. c. novaeguinea:* Mid-mountain valleys of New Guinea.
- *C. c. papuensis:* South-eastern New Guinea.
- *C. c. colletti:* Northern Territory of Australia.
- *C. c. australis:* Queensland to Victoria.

It is difficult to know which of the above subspecies has been imported to make up the many birds that are now in the pens of United States aviculturists. More than likely, the known nominate race *(C. c. chinensis)* make up the bulk of the captive population and from time to time importations of other subspecies probably have been made. The many color mutations now available from Australia and Europe are of unknown subspecies.

A Close Relative

I should mention the African Painted Quail which is the only close relative to the common Chinese Painted Quail as some readers might encounter them. This would more than likely never happen under the current circumstances. However, recently a bird importer told me of two males getting into a consignment that he received from an african trapper. They made it through quarantine but later died of an unknown cause. It is a shame that this was not a pair and they had lived to reproduce in captivity! If they should ever become available, I certainly would love to have some. At the present, I am not aware of any that are in America. Some effort is being made to locate and import this interesting species. Most of the literature describing this quail states that they are rare in the wild. This sad fact would just about preclude any ever becoming available to the aviculturist in America in any numbers. I have personally questioned several bird importers and they have never run across this bird. Also, personal correspondence with a top breeder in Europe has indicated that he knows of none in captivity.

They are called by the same names with the addition of the word "African". Their scientific name is *Coturnix adonsonii* and are found rarely in their large range. A distinct color difference is seen between the Asian and African birds. The latter males lack the chestnut coloration on the breast and underparts. The females have more black barring on the wing coverts (Johnsgard, 1988).

I often have callers inquiring about the possibility of getting these birds imported to the United States. Very few trappers in Africa are able to catch them and few of them even know what they are. If any of these birds

27

should ever be available to aviculturists they should be easily propagated in the same manner as their cousins from Asia.

No doubt special care should be given to newly imported birds as they will be skittish for a few months until they are acclimatized. Give the newcomers as much privacy as possible and disturb them only as much as is necessary for their care.

Five week old Normal and Silver mutations in the growing pen.

Description

Measuring only 4 1/2 inches, the Chinese Painted Quail is the smallest of all the quail found in the Old and New World. Its size alone makes it very appealing to nearly everyone that sees one. This is especially true when the fortunate have the opportunity to view and raise the tiny chicks. They are smaller than the thumbnail. It is truly amazing to see newly hatched chicks running around so full of energy. They are able to fly in just a few days after hatching.

"The cock is the brightly colored one in the family. His breast is covered with blue-gray and burnt-red markings. His face has a blackish area which is outlined by a black line extending from the beak, below the eyes and join the black patch on his throat. It has a broad white crescent below the black throat extending back to the ear coverts bordered with black. He is a beautiful little bird and must be examined closely to appreciate the colorful markings.

The hen is mottled brown and lacks the black and white markings. She is rather dull in appearance compared to the cock, but her markings are subtle and very beautiful upon close examination. Her under parts are not as dark a brown as is her upper parts" (Rutgers, 1965).

All of the literature about the Chinese Painted Quail in the wild is limited. Practically all of the information we have on record is obtained from captive bred experiences. Field studies are hard to accomplish because of time and expense. Very little has been done toward getting information about this bird in the wild.

"In the wild, this little quail is scarce. It lives in the scrub jungle along the edges of marshlands" (Fleming and Fleming, Bangdel, 1979).

Understanding Chinese Painted Quail

These quail have some bad habits shared with other members of the Coturnix family. I must say there are many more good traits of these little birds than bad ones. Do not think that Chinese Painted Quail are the only game bird that has bad habits. All wild birds held in captivity exhibit peculiarities that must be dealt with by their keeper. The finest solution to these problems is to avoid them if possible. The accomplished breeder will be very aware of the following bad traits. The new breeder will be well acquainted with them after a season of experience. It is hopeless to attempt to change these traits. The wise breeder will learn to adjust to his birds, rather than expect the birds to adjust to the conditions the breeder decrees.

Wild Flight

Most Chinese Painted Quail have the trait of *flying straight up* and hitting the top of the pen. We have had all kinds of disasters from this obnoxious characteristic. When this first happened, I began to wonder what I had gotten myself into. On our first night with our newly arrived Chinese Painted Quail they began to fly up and hit the top of the pen constantly in panic. Even after turning on the lights it took a few minutes for them to calm down. Luckily, no damage was done other than scalped heads which soon healed, but we learned an important lesson in handling these birds. One way to control this habit is to keep their wings clipped, especially if they are in a tall pen. This will need to be done every so often during the growing period.

Another thing to do is to put a soft material in the top of their pen so when they hit it the surface will not be hard enough to damage skulls. This can be fiberboard or even a piece of canvas that is hung by the corners next to the top of the pen. Be careful if you use styrofoam as some of the birds may bite off the material and eat it. (We have never lost a bird from ingesting styrofoam, but it is best not to take any chances.)

Male Aggression

Another common fault--the males have belligerence to other Chinese Painted Quail males. This is really not a fault, but simply means the male has a strong protective nature towards his family. This is seen in several species of birds.

We learned that the male is quite timid compared to his mate. The female will be happy to come to the fingers for a mealworm but not the male. Sometimes, the hen will take the mealworm into the hiding place to share with the male. This is a characteristic of this family of quail that will become treasured by the observant breeder.

The real thrill comes when the hen or the cock will call out the babies to take a mealworm or bit of food from their bills. The little ones rush out and grab the food with such glee that you know you are seeing something few breeders see. This is a real treat to see.

> HINT: Be careful if male Chinese Painted Quail
> are kept together during breeding season.

Cannibalism

Cannibalism is also known as *toe picking or feather picking*. Some Chinese Painted Quail which are kept in very small pens develope this habit. Picking the toes can kill or cripple young chicks. Even though some injured chicks will survive, they are a sad sight to see.

Several things can be done to prevent this problem before it starts. Once started, this habit is hard to break and the offender must be debeaked or isolated. If we can determine the cause of the problem, it can be prevented. Some strains of Chinese Painted Quail are extra bad about this while others are not bad at all. Even different hatches of the same strain can develop into pickers. The truth is that we do not know the cause of this habit. It could be one thing one time, and another the next, so any general conclusions could be misleading. My advice for the new breeder is to try the different methods and stick with the one that works.

We put stemmy alfalfa hay in the bottoms of the pens which seems to discourage toe picking. The hay stems break up the outline of the toes and discourages picking. Many of the photographs in this book show the use of stemmy alfalfa hay. The bird droppings fall through the hay and wire and most of the time the birds live happily ever after. Red lights and a dark box seem to help also. Never put breeding birds in a darkened area as they will go out of breeding condition. Also, be sure there is enough light for the birds to see to eat and drink. Every so often we hang up some lettuce and let the birds pick at it. (They like lettuce with salt lightly sprinkled on it.)

Chicks often develope the bad habit of feather picking. The soft quills which are filled with blood are especially appetizing to the chick that picks one of his fellows. Small feathers on the back and wings are most easily accessible. Feather picking many times occurs around the vent and tail area. Once blood is drawn from a bird, its companions go to work picking blood and meat from the victim. Unless the picked bird is removed and treated, it will be picked to death. A very bad habit will be started. Feather picking, once started is very difficult to prevent (Wilson, 1972).

To help conquer this gruesome obstacle, we have used a product that is a red paste which prevents feather

> HINT: Keep trying various cannibalism controls
> until you find one that works.

picking. It seemed to help in most cases. Isolating the
individual picker is another solution.

Along with the electric fence charger, the most
valuable piece of equipment for us through the years has
been the "Debeaker" which is manufactured and sold by
Lyon Electric (See advertisements) and has paid for
itself a thousand times over. We actually do not debeak
our birds but rather just "sear" the sharp tips of their
beaks. This has to be done every two weeks or so on
growing birds, but the advantage to me is that when the
birds are mature they aren't disfigured by the debeaking
process. This operation serves as a deterrent during the
sensitive growing season. Many of the birds will
outgrow this tendency if they are not crowded or otherwise subjected to an environment that encourages this
bad habit.

One must resolve that the so called causative factors
related to picking are controversial and complex and
really not understood. Cannibalism may occur under
the most promising conditions and not under less

advantageous conditions. It is a mystery why it occurs in some broods and not in others. I have had birds (from the same bloodline) be terribly cannibalistic while their siblings never give any problems.

The Lyon Electric Debeaker is a valuable piece of equipment.

Some good very good *"tips"* for cannibalism prevention are (Woodard, 1977):

❑ Provide adequate floor/pen space for the birds.
❑ Remove dead, sick, or weak birds immediately.
❑ Remove obstacles that may cause injury.
❑ Never introduce birds during breeding season.
❑ Restrict human traffic near bird facilities.
❑ Provide adequate feeder space and waterers.
❑ Avoid sudden changes in texture of feed.
❑ Avoid sudden changes in temperature.
❑ Provide adequate shelter, cover, etc.
❑ Use good methods to control cannibalism.

HINT: Do your best to keep cannibalism from starting. Prevention is much better than trying to find a cure.

An Interesting Experience

We discovered that these little birds have a kind of instinctive hatching language. They call to one another, and when one hatches there will no doubt soon be others. The chicks do not cheep like chickens but have a strange little call that says *"eeeh-eeeh-eeeh."* Their call is

more easily heard when they are hatched. It is be-
witching to watch them just when an egg hatches. Many
begin to hatch within a few minutes of each other. This
hatching observation is alone is worth keeping them. I
am not totally persuaded that they can regulate their
hatching time.

Very Good Neighbors

A most delightful characteristic of this little bird is
the ability to live happily with other species of birds. It
is totally non-aggressive to others (except its own kind
during mating season). Chinese Painted Quail can be
kept in the bottom of an aviary or cage with other types
of birds living in the top area of the cage. The ground
feeding quail offer no competition for food that is kept
in seed containers for the top-dwellers.

Many cage-bird keepers have a pair of Chinese
Painted Quail in their aviaries which pick up spilt seed.
This practice adds much to the joys of aviculture. There
is nothing more rewarding than to have an aviary of
mixed compatible birds living and raising their young
harmoniously with their neighbors. This can be
accomplished if careful planning goes into the choosing
of the occupants of the aviary.

The noted authority on Finches, Robert G. Black, says the following about the value of Chinese Painted Quail in the Finch aviary:

"Chinese Painted Quail can be extremely valuable in an aviary as natural brooders for finches that have left the nest a little too early. A pair or trio is best, for two males kept together will fight to the death unless they've been raised together from the chick stage. Though I have never seen them harm a finch, the Chinese Painted Quail are utterly vicious toward one another. Their great value lies in their habit of dwelling on the ground where young finches may be. On cold nights, the fledgling finches will crawl under the quail and be completely warm and comfortable for the night, as well as during the day when the quail are quiet. I've seen many finches brooded and protected in this way even on sub-freezing nights. Fortunately, the fledgling finches instinctively seek the brooding quail when the temperature begins to drop. I've frequently observed a female Chinese Painted Quail sitting on her own eggs with young finches peering out from beneath her feathers" (Black, 1984).

We soon learned that Chinese Painted Quail could not tolerate very cold weather unless protection is provided. Always provide open drinking water and a place to get out of the wind. To be safe, give them a place to get that has a solid bottom such as a board for them to rest on. Standing on wire in cold and windy weather can be very hard on them. It is best to keep them where the temperature stays above freezing. They can tolerate sub-zero weather only if adequate shelter is provided. Most people do not realize that these little birds are tough if they are hardened to the weather over time. Do not put them directly out in the cold as they will chill and get sick. Let them gradually adjust to the change in temperature over a period of a few weeks.

Many years ago we first had these beautiful little birds and tried to keep them in a large outside aviary in Montana. A good shelter was provided to let them get out of the wind and snow. They had no trouble with the snow but were killed by some Bobwhites that were sharing the pen. This was a heartbreaking but educational experience for us, one which we will never forget!

Calls of the Wild

The calls of these birds are low and will not disturb any neighbors. Most of the time your neighbors will not know that you have Chinese Painted Quail in your garage or back yard. The adults make a cute little sound described as *"tir-tir-tir."* The males also have a piping whistle and a crowing sound during the breeding season. It is certainly not loud enough to create any disturbance.

Chinese Painted Quail have a very interesting communication system. Years ago we had Chinese Painted Quail, but they were kept outside and we never heard any of their calls. Only when we were around them more did we learn of their many calls. We discovered this system of calls really quite by accident. We got our breeders in the middle of winter and had to keep them in one side of the garage because of cooler temperature. My study and den has a connecting door leading into the garage. To our amazement the birds began to call to one another. They would answer with such enthusiasm that we soon discovered that they had a certain language.

Here would be a good place to play a recording of their calls so that the reader (listener) could understand just exactly how they sound. (I am sure that somewhere, someone has done a doctoral thesis on this part of Chinese Painted Quail behavior.)

Later we put our brooder boxes in my office and found that even very young chicks have the calls well developed as they would call to the birds out in the garage which was only a few feet away. The unmated males seemed to call more than the pairs as they were wanting to mate with some good-looking quail of the opposite sex.

The most amazing of all of the calls was the noise made by un-mated mature males. They would expand their throats and make a hissing, growling, raspy noise similar to the moaning that members of the dove and pigeon family make. This activity stopped as soon as the hen accepted them so apparently this is a way to court the females into entering a pair bond. The extra males housed together continued to make this call.

I just cannot talk too much about the added enjoyment we got by listening to these little birds call. It made the pleasure of keeping them much richer. By all means, keep your Chinese Painted Quail in an area that you can listen in on their conversations which will add to the enjoyment of these little creatures.

Chinese Painted Quail Postures

Several studies have been made to determine the non-verbal communications of Chinese Painted Quail. A good understanding of these will help the breeder in the enjoyment of his quail. Simply by watching your birds you can determine much about their general well-being. Schleidt (1984) recognized at least 54 of these postures with variations (Johnsgard, 1988).

A Silver mutation male (center) showing the white markings.

What Is A Color Mutation?

When many thousands of birds are bred, sometimes there is a deviation from the normal gene structure which sets color. These are called *sports or mutes*. The normal color patterns are changed and thus new colors appear.

The most popular color mutation is a **Silver Mutation**. These birds are very beautiful. The silver color almost covers up the pattern around the neck but it still shows very faintly. This makes them strikingly beautiful. Also, the silver coloration is much brighter than the silver coloration of **Silver Bobwhites**. Silver mutation in other birds such as pigeons are described as *powder silver*. We have had Silver Chinese Painted Quail and like them very much. They are certainly not as colorful as the normal colored birds. They are still very beautiful indeed. The price of the Silver Chinese Painted Quail is sometimes slightly higher than the more common normals, but some breeders charge the same price as they charge for the normals.

There are several new mutations now available in America, Europe and Australia. One of the first available mutation is the **White Button Painted Quail**. These are beautiful little birds. They have dark eyes which show that they are not albinos. Their legs and beaks are bright orange and their feathering is pure white. No color pattern shows at all which sets them apart from any other mutation now presently available. The way to sex mature birds is to look for the larger plumper birds which are usually females. This is especially true if they are currently laying eggs. The males are somewhat smaller and more trim than most females. The chicks have bright yellow down.

44

Although at the present these white mutations are rather expensive, the price will come down as they become more available. In all fairness, the original importers have to charge more for the birds to *"re-coup"* their extremely high investment in bringing in the birds through quarantine.

Now available on the market is a pretty mutation called the **Red Breasted Button Quail** which was developed in the United States by Garrie Landry of Franklin, Louisiana (See advertisements). It is now being bred and is available from breeders. It's color is about what the name implies it should be. It appears that this color mutation will perhaps one day be as common as the silver depending on the likes and dislikes of aviculturists.

The females have black barring from head to tail over a buff-tan foundation coloration. The males are similar to normal colored males except they have no white markings on their throats. Their face is black and the breast is completely red from the throat to the vent. About one-half of the wing feathers are red which makes them appear half red, half black when viewed from the side or front. The chicks are distinctly marked also. They are dark above and yellow below. Many agree that these are a most outstanding color variety of Chinese Painted Quail.

The fact that this is a new mutation will no doubt make it popular with some aviculturists. It remains to be seen if this mutation will catch on and be as popular as the Silver Chinese Painted Quail mutation.

Some breeders are now propagating in large numbers a **Red Breasted Silver Mutation**. The hens have cream-buff background with dark silver barring from head to tail. The males have a silver appearance with a grey-silver face with no white on the throat. Their breast is delicate subtle pink from the throat to the vent. The wings also have a pink coloration. I believe that these *"silvers"* are much more beautifully colored than the common Silver Chinese Painted Quail although they are somewhat more expensive. Their price will go down as more and more breeders raise them.

A mutation to arrive recently in the United States has been imported (Fall of 1989) by Garrie Landry. It is called a **Fawn Mutation** which was developed in Australia. It has come into the states via Europe through a U.S. Quarantine Station. The color is said to be "*cinnamon*" and very pretty.

There is also available a color mutation that is called **Blue Faced Button**. At this writing I have not seen this color but it is described as having the entire breast, belly, face and forehead blue. The blue completely eliminates the facial pattern in the male. The backs of the males are dark brown. The beak is jet black and males actually resemble rails rather than Chinese Painted Quail because of the dark appearance. The female of this mutation is dark chocolate brown. The chicks are very distinguishable as they are very dark chocolate brown and rust red.

The Red Breasted mutation is beautiful and has become very popular.

A beautiful Silver hen showing the barring on the breast and the streaks on her back. This is the most popular mutation.

The **Blue Faced Cinnamon** is a combination of the Cinnamon and the Blue Face mutation. It is very attractive. This is a new mutation at this writing and only one breeder has these in his aviaries.

The **Golden Pearl** mutation has only recently came into the United States out of Belgium. The hens have less brown pigment, each feather is straw yellow with light brown barring. These hens are certainly the most colorful of all Chinese Painted Quails. Golden Pearl males could be mistaken for normals. This variety resemble the Manchurian Golden variety of the Coturnix Quail which was developed in 1960 through 1965. Pearl chicks have black and yellow down.

The **Cinnamon Pearl** mutation brightens the cinnamon mutation at its best. These birds have bright orange color throughout. The crown is very bright, and the body color is mostly light orange with grey barring. The chicks are quite colorful and distinctive.

The **Ivory** mutation is another new variety just imported. The birds are a pale ivory-white all over with no grey markings on the feathers. They are much lighter than silver but not pure white. Males show a prominent blue-grey breast on an otherwise ivory body. Hens are all ivory with no silvery grey plumage.

The **Red Breasted Cinnamon** is being produced presently by Garrie Landry and should be available for sale in 1991. The bright red breast is very prominent.

I have heard from Garrie Landry about a **Melanistic** (black) mutation which he is now breeding. These birds are distinctly different and are mostly black in color. It should take a few more generations to get the solid black to breed true. He is getting a few *Black* birds and is planning to breed from them. If this can be done, this will be the latest of many more mutations that should be developing over the next years.

They are coal black with black eyes and orange legs and feet. As far as I know, these are the only birds like this in the United States. I have heard rumors that some are now being bred in Europe (but no confirmation). They should be very popular especially with Chinese Painted Quail breeders that are real enthusiasts.

HINT: To keep strong stock, it is good to out-breed the mutation with a normal every fourth generation or so.

To study the differences, between the normals and the mutations be sure and have plenty of space. Many times breeders get all of the different colors and then choose one or two that they like the best. To keep unrelated birds in each color would mean that lots of birds would be involved. This would be a real challenge to the dedicated quail aviculturist.

As time goes on, there will be as many mutations with these birds as there are with the Bobwhite Quail. We can look forward to other very interesting color mutations in the coming years.

We have had some of the above mutations in our pens while doing the research for this book. We do not keep Chinese Painted Quail at the present time. Everyone has favorites and I am hard pressed to give you the color mutation that is my favorite.

I have changed my mind several times during the research for this book. Right now, I must say that my favorite is the Normals. It is difficult to improve on nature and these little birds that nature gives us are so beautiful. There is something exciting to me in keeping all my birds as "pure and close as possible to the wild." This may sound naive, but I really feel this way. There are thousands of game bird breeders that feel the same way.

While on the subject, it should be remembered that color mutations in these quail are not gotten by cross breeding with another species of quail. They are pure Chinese Painted Quail with no blood of another species in their veins. I am personally not in favor of this practice but I know lots of people that see no harm in doing this.

HINT: Some system of keeping the eggs of the various mutations separate should be used.

In the incubator we use individual little baskets made from 1/8 inch hardware cloth. Lids on the baskets keep the newly hatched chicks from jumping out. Since each of the mutations throw different colored chicks they can be easily identified.

When the little chicks are removed from the individual baskets (all the same bloodline), we use a "toe clipping system for permanent identification. The system is as follows:

Left hind toe clipped - Bloodline #1
Right hind toe clipped - Bloodline #2
Both hind toes clipped - Bloodline #3
No hind toes clipped - Bloodline #4

This system allows four bloodlines per mutation. This should be plenty unless many chicks are hatched.

CHAPTER 2

BUYING
CHINESE PAINTED
QUAIL

The person who brings home any pet should have already made the determination to take on a serious responsibility. The person who does not have any other birds should make sure that he or she understands what will be required.

A real aviculturist would never be happy without birds. Others would never be happy with birds. Sooner or later the decision must be made whether or not the *person will keep birds*, or if the *birds will keep the person.* It is true we can become prisoners of our feathered friends. Love of birds can be a demanding mistress. This situation has happened more than once and I know many breeders that have overloaded themselves and their pens with too many birds. Most of the time when the breeder only keeps a few birds, real pleasure is experienced and valuable information can be obtained through careful research. It is a simple matter of how much time can be given to the care and satisfaction of the birds. I am speaking of personal experience and give this warning to those that will listen.

I do not want to dishearten the reader from getting Chinese Painted Quail. If anything, my purpose is to encourage those *"bird lovers"* to get a few. I suppose that the motivation of the keeper is as much of a consideration as anything. Why does one want to keep birds? If the answer seems satisfactory, then by all means get some Chinese Painted Quail. These little birds will fit into most any lifestyle pattern that you may have. They only require the necessities of life.

HINT: If possible visit the breeder to inspect his premises. Pick up the birds to save expensive shipping costs.

Three Places To Buy

There are several ways to obtain Chinese Painted Quail. I mention three initial ways you may get them.

FIRST, you can buy them from a local pet store or in the pet section of a large department store. This is perhaps most often done as most pet shops of any size carry these little quail.

HINT: Have a good look at the store. Are the cages large enough and clean? Do the birds look healthy? If not, go to another store!

SECOND, you may get your start from a breeder who has surplus birds for sale. This is often hard because there are not that many local breeders that have Chinese Painted Quail.

THIRD, you may order them from a supplier out of town or state. This is perhaps the most risky way of doing it. The birds nor their surroundings can be examined ahead of time. The buyer must take what is sent sight unseen. Regardless, we have done this for years with no problems. (We always try to buy from reputable people who provide references.)

Select birds that are energetic and not standing in a corner fluffed up. If they are busy picking up seeds, they are no doubt in good health. Have the clerk catch up a prospect and look closely at the eyes. Each eye should be clear and bright and not have any discharge or drainage. Feel the breast-bone to see if it is thin. Examine the feathers for tightness. Pick out birds that are fully feathered and not picked. The clerk will insist that the feathers will grow back and this is true. If a feather picker is bought, you will bring home problems. Be convinced that you have a true pair as two males will fight and two females will only lay infertile eggs.

> HINT: Be sure and find out what **guarantee** the store offers. Get it in writing if possible.

Arriving in the New Home

Be certain and have everything ready for the birds before you bring them home. When you get your new charges home, keep them as quiet as possible in a darkened area for a day or two. Be sure they have enough light to see to eat and drink, but the rest will do them worlds of good. Resist the temptation to constantly look at them. Do keep an eye on them to be sure everything is going alright, but do not unduly disturb them. On of the hardest things we must do is to leave our newly acquired quail to themselves. There will be plenty of time to get to know them later when they have settled down to their new home.

The birds should stay in the shipping boxes only as long as required. I have had Chinese Painted Quail sent by air across the land with no problem. We always put small pieces of apples and oranges and some feed in the bottom of the box for them to munch on when we ship birds. They do quite well but the longer they must stay in the box, the more stressed they become.

Chinese Painted Quail ship very well when they are mature. Do not ship very young birds as they are fragile and may not make the trip. It is a good idea to only ship birds at least three months old or older. Avoid shipping in weather that has temperature extremes. Very hot or very cold weather tends to stress the birds. Many airline companies will not ship if the temperature gets out of a calculated range as they have lost a lot of birds when they ship under these extreme conditions.

The new Express Mail for chicks now handles quail if properly boxed.

CHAPTER 3

BASIC HOUSING AND EQUIPMENT

The decision must be made as to what kind of surroundings will be provided for the Chinese Painted Quail. If kept in the bottom of pens with other birds, there is not much extra care. Provide food and water and other requirements accessible from ground level. One pair can easily be kept per pen; more if precautions are taken.

Most cage bird people (those that raise parrots, finches, etc.) keep a pair of Chinese Painted Quail in the bottoms of their cages. Be sure that the birds get along. Dick Schroeder tells us about a disaster he experienced with Chattering Lories and Chinese Painted Quail.

"Several years ago we housed our first pair of Chattering Lories in a 10-foot-square cage. This wonderful flight cage had a tree growing in the center. We added a pair of Chinese Painted Quail as grounds-keepers. All the

birds got along well...for a while. The quail hatched several chicks, which ran around the floor with the parents. One afternoon the Chattering Lories neatly nipped the heads off of all the quail, including the adults" (Schroeder, 1989).

These little birds get along just fine living with many of the non-aggressive finches. One should always be careful to see that the birds find the food and water founts.

Some breeders like to keep their Chinese Painted Quail in specially built private breeding units. These measure about 9 inches by 9 inches by 12 inches, each of which can hold a pair or trio of birds. The bottom of the cube has a slanting floor which permits the eggs to roll out onto a ledge. This way there is no disturbance when gathering the eggs. There are commercial supply houses that sell these cages in units which can be stacked. If kept in these close quarters during the breeding season the birds should be rested a few months each year so as not to loose their stamina.

> HINT: Design your pens so you will need to disturb the occupants as little as possible during breeding season.

If you are good at building things you may want to build a nice outdoor aviary. You can keep the quail outside during the Summer months if you live in a cold climate. In more temperate climates, Chinese Painted Quail can stay outside all year long with no problems as long as they have the needed shelter during storms. Let your imagination go wild. Keep in mind that wet areas breed bacteria and thus are problem areas to avoid. Plan to keep the bottom of the pen as dry as possible. When you have plants that need watering, water only enough to keep the plants in good condition. Never let water stand in puddles. I like to build my outside aviaries with a layer of sand that is above ground level to avoid standing water. Besides giving good drainage, this arrangement permits the sand to be removed once or twice a year completely and replaced with clean material. Also, the sand can be raked once a week and the droppings discarded.

Of course, if the aviary is outside it must be protected from predators. A simple lock on the door is needed. Skunks, weasels, dogs, raccoons, coyotes, owls, hawks, and cats are some of the predators to consider.

Many years ago, we found that the best deterrent to predators was an ordinary electric fence charger. You can buy these in any farm store. For about $50 you can have a charger that will last at least 10 years which is a very prudent investment. We run a hot wire about 5 inches above the ground along the bottom of the outside walls. This keeps the diggers out. We also run a connected hot wire along the top of the outside walls about 5 inches above the pen. This charged wire keeps the cats from walking and owls from landing on the pen.

Much care is needed to place the electric fence so children and pets cannot get into it. A regular fence a few feet around the quail pen will deter intruders. Besides giving a bad shock, under certain circumstances the "hot wire" could be dangerous. Place warning signs every few feet. These are bought in farm supply stores and they are brightly colored to stand out.

Now for a word about whether or not to cover the top of the aviary. By all means plan to have a covered wire top. I have found the best material to use is the new plastic netting for the top of the pens. It is much easier to install and saves the birds when they fly against it.

Using open top pens with the small Chinese Painted Quail is asking for tragedy. Use the small 1/8 inch hardware cloth or special aviary mesh for the sides. This is especially important if you plan to let the quail brood and raise their own young. The tiny babies can go through the normal 1/4 mesh. The top can be made

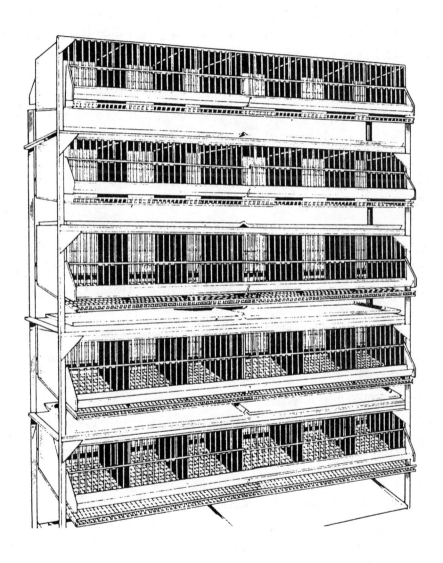

Breeders that want to raise large numbers of quail can use a stacked breeding unit such as this.

of the larger 1/4 inch mesh as the quail will not attempt to get out through the top. If small finches are kept in the aviary, place the small sized mesh over the top and all sides of the pen.

> HINT: Have the open area of your aviary designed so that you can see the birds from your house through a window.

If you plan to build a walk-in aviary make it at least 6 feet high or higher. If you plan to put in birds to occupy the top space make the pen as tall as possible.

Years ago I built my pen just 5 feet high. Since I am over 6 feet tall my back is still sore from all of the bending I had to do. Also, make the doorways wide enough to get a wheel barrow through easily. There should be some form of shelter at one end. This shelter is backed against the prevailing winter winds to give some protection during storms. Pieces of plywood work just fine but should be made waterproof. It would be a good idea to have the shelter to cover about 1/3 of the pen area.

Nothing is more pleasant than to sit in the backyard in the cool of the evening and watch your birds. Locate the outside aviary where you can see it yet put it as far away as you can from your neighbors. Try to keep day and night disturbances at a minimum by carefully locating your pens. Avoid placing the pens where automobile lights shine into them at night.

Spanish architecture with the famous *"patios"* can be designed to accommodate a very nice aviary which will be in full view of the house. Attractive pens for our birds does much to win friends and influence our neighbors.

I am currently planning to build some interesting pens. They are constructed from a series of standard panels which are bolted together. Many configurations can be assembled using the panels. The system can easily be enlarged (Larosa, 1973).

All of us dream of having a special house in our backyard just for our birds. If this is possible it should be as large as we can afford. A one room house would work beautifully. The house should be well insulated. It should have ample light and ventilation. Some kind of heating unit should be used if the temperature drops below freezing.

Some of the more liberal bird breeders can transform an unused room into a bird room. We have done this on several occasions and it worked for a while. Soon we tired of the dust and extra work involved in cleaning. Too, our neighbors began to wonder about us! We were very glad it was a temporary housing situation. We found out that we are slightly allergic to feather dust which is bad when the birds are kept indoors. The ideal place is what best works for you. The climate conditions, security considerations, and the amount of money that can be spent will dictate the housing for your birds. However, let me emphasize again that pens do not have to be elaborate or expensive at all. These little quail will live as happily in a shanty as in a mansion. They never complain (at least in English).

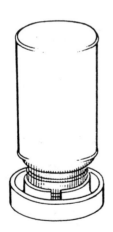 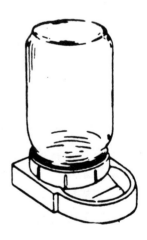

Always get the special small quail waterers as the size for poultry chicks are too high for the tiny quail chicks to reach.

Some Needed Equipment

The amount and type of equipment that the prospective Chinese Painted Quail breeder needs depends on how many breeder birds he has and how many young he wishes to produce. If only a pair or two are kept, there is no need to try to incubate the eggs artificially. Just let the birds make their own nest and raise their young naturally.

Use the proper size feeders and waterers. It is important to have the small size especially designed for small quail. Many supply houses have these that are made out of plastic that can be washed quite easily. We use the plastic lids off of margarine cartons to put feed in when the chicks are very young. Mason jar lids filled with marbles or pebbles make good waterers if the small commercial ones are not available.

HINT: Be careful to keep the plastic from very cold weather as it will become brittle and crack.

Dusting boxes are a must for Chinese Painted Quail that are kept on wire flooring. These can be made of wood or other material. We have used one gallon milk containers cut off filled with sand quite successfully. If the birds are kept on dirt floors of course there is no need to have these.

Grit boxes provide the needed trace minerals the birds need. These can be filled with sharp river sand, charcoal, and even the special mineral mix that is manufactured for pigeons.

Nest boxes should be used in wire cages. Many times they will be ignored and the quail will lay on the wire floor. Cut milk cartons work for these also.

The Bird Net is one of the best friends a breeder has. A good net will save lots of accidents in catching your birds. Get a net that has a long enough handle to reach the back corner of the pen.

A net used for catching the birds is very important to avoid injury to the birds. Get a net that is deep enough and large enough to easily catch the birds. It should be small enough to go through the doors of the pens.

It is a good idea to have several spare pens available. These can be used when moving birds or when cleaning out pens, etc. We have also used large brown paper bags for this from the grocery store. Simply cut air holes and put a handful of straw in the bag for solid footing and the birds do just fine for the short time they are in the bag. You may have to request this type of grocery bag these days as they are being replaced by the plastic kind. Do not use the new plastic grocery bags as the birds will suffocate.

Other items that will come in handy are cutting pliers (for wire), scissors, first-aid materials, disinfectants and other medications. The breeder will have a lot of necessary items on hand that can be used for the birds.

If many birds are to be raised it will be a good idea to get an incubator. This may seem to be a large up front investment for the small breeder, but in the long run an incubator is one of the most valuable pieces of equipment.

This homemade brooder box holds chicks up to four weeks of age.

The growing pen can easily be constructed by the handyman from plywood.

CHAPTER 4

BASIC CARE OF CHINESE PAINTED QUAIL

Whatever we do, we always do it better if we have some "routine." This is undeniably valid in caring for Chinese Painted Quail. What I am suggesting is to put into practice a word known in the business world as "standardization." This simply means that you should develope some systemized way to care for your birds.

Your system will be unlike mine. I will include some things that you will not embrace. Both of us will be right and neither of us will be wrong as both of our systems get the job done. If there is any one thing that successful game bird breeders have in common it is "they care for their birds well in a systematic manner". This is especially the case when a large number of birds are raised. Also, if there are a number of age groups to care for a system is a must as feeding requirements vary.

Some of the many things to put into a standardized check system could be:

☐ The time of day the birds are cared for.
☐ The person who takes care of them.
☐ The order that each pen gets feed and water.
☐ Other maintenance chores.
☐ The diet and the treats that will be fed.

There are untold things that could be put on the list. The more birds you have, the more you need a standardized system. Develope some system that helps you get the job done at the precise time. This is especially valuable for people like myself that have trouble remembering if I have done a specific chore.

Feeding Chinese Painted Quail

A main concern is to get our Chinese Painted Quail on a well-balanced diet. This can be done regardless of their pen arrangement if we use some common sense. If your Chinese Painted Quail are kept in the bottom of a parrot aviary, their diet should by all means be supplemented so that they get more to eat than just seeds.

Have some commercial game bird feed available for them in little containers on the floor. The extra nutrients will give a good balanced diet.

> HINT: It is important to get your birds to eat a variety of foods to capture the trace elements that they need.

Chinese Painted Quail can be raised totally on commercial game bird feed. Do not feed them a ration prepared for laying hens or chickens. There is not enough protein in these feeds. Chicken laying mash or pellets has too much calcium which counteracts some vitamin elements. However, we have fed exclusively a commercial Turkey Feed to our Chinese Painted Quail and have had good success. The nutritional requirements for Turkeys is very near the same for Chinese Painted Quail. By the way, do not buy 50 pounds of feed for a pair or two of Chinese Painted Quail. It will take too long to feed it up and it will loose much of its nutrient and vitamin content from long storage. Get the feed store to sell you a few pounds at a time so as to always keep fresh feed available.

I was asked just this week to tell the caller any secrets that I might have in raising quail. I told the gentleman that the one thing that I believe makes the birds lay is a supply of fresh grass during the winter months. Most people know about the *"winter grass"* that

73

grows in early spring that is so green and luscious. Simply pull this up in clumps and throw the roots, dirt and all in the pens for the birds to pick at. This not only gives them something to do but they get trace minerals from the soil. If I have a secret, this is it!

HINT: Never feed old or mouldy feed.

Some "Extras"

There are some *"extras"* that can be done which will help keep Chinese Painted Quail in good physical condition.

One of our habits that we enjoy is feeding our birds seed by hand every day. This gives us a good chance to look them over for problems, and it also gets us involved with them in a more personal way. The extra seed during periods of stress such as cold weather gives them the needed body heat to get through cold winter months in an outside pen. Of course this cannot be done if the quail are kept in the bottom of hook bill pens. In this case feed some food that the quail especially like and you will get the same results.

Feeding treats by hand daily gives me a chance to study the habits of the birds. When they come to the hand for feed I can observe them easily from a different perspective.

Tame birds show off better to visitors. Our birds come over to the wire fence to get their treats. I tell our visitors that they love me and want to be near me. Of course, soon the truth comes out that the birds really love the feed.

We have always had the habit of giving nutritional extras to our birds. These keep our birds healthy and happy and can include various vegetables from our garden. The birds just love them and look forward to getting them daily. (Another thing that most quail like is apples). Chopped carrots from the garden are readily eaten by Chinese Painted Quail if they are shredded very fine or chopped into small pea-sized bits.

Give the birds as much *"green foods"* (fresh fruit and vegetables) as they will eat. Not only does this give them the extra nourishment, but insures they get needed trace elements and vitamins the natural way. Artificial minerals are difficult to balance, stay with the natural if possible.

Feeding **bird cornbread** is another way breeders can insure their birds get proper nutrition. We have no idea who originated this recipe for bird corn bread. Bernard Teunissen gives us his recipe. He uses four-sided cookie sheets in preference to baking pans, as this makes a dryer product and less prone to spoilage. The bread can be cut up and fed in pans or crumbled directly on the feed in the feeders. Both ways provide extra nutrition for adults and chicks alike.

The recipe: *Mix 3 cups small bird seed (such as millet); 2 cups each of whole wheat flour, soya meal, cornmeal; 1 cup Vionate (vitamins); and 1 cup baking powder (yes, 1 cup! I tried it without, and it was like a rock); and 4 teaspoons salt.*

You mix all these dry ingredients and then make a "well," and add 5 cups eggs (shells included); 1 cup milk; and 1/2 cup wheat germ oil. Then bake at 425 degrees in 2 cookie sheets, 12 by 16 inches, for about 25 minutes or until firm to the touch. I use Teflon coated pans.

The bird seed is put in basically to give the bread body, otherwise the bread will be tough. When feeding the formula to non-breeding birds it is best to replace the wheat germ oil with plain cooking oil since the bread need not be as rich and oily (Gazette Editors, 1978).

Most of the time our birds take readily to these extra foods. However, we sometimes get one or two that need to be helped along. Once the timid ones see how much the others enjoy the new food they join in. It is important to get as much variety into the diet as possible. Be sure not to just have spilt bird seed as a regular diet. Supplement it with some kind of high protein foods.

Vitamins and Minerals

A very important nutritional need is **vitamins and minerals** which in most instances are lacking in Chinese Painted Quail diets. When birds are kept on wire they cannot get trace minerals or vitamins from the soil. To help overcome this problem, we add commercially prepared vitamins to the diet. Some breeders sprinkle a powdered vitamin product, Vionate, over the feed in the troughs. Vionate can be put on treats. This is an excellent product and we heartily recommend its regular use.

We were amazed to find that Chinese Painted Quail chicks of all ages just love to eat the Vitamin Supplement called Vionate right out of the bottle spread on the floor or in a flat lid. They seemed to like it so much each day we would sprinkle about one/half teaspoon of this product in each quail box. This would be enough for around a dozen chicks. Order a can of this wonderful product (see advertisements).

We also sometimes use a water soluble vitamin product and keep it in the water. The problem with this product is during warm weather it can become a breeding place for bacteria and must be changed at least everyday. You may prefer to put vitamins in the water every other day.

> HINT: Vionate can be ordered from Gazette Medications Department, 1155 E. 4780 So., Salt Lake City, UT 84117. (See advertisements)

When deciding on a vitamin product you should be sure that it is formulated for "birds" and not for other livestock. This is why we use Vionate as it is specifically formulated to meet the needs of birds. It can be ordered through the *Gazette* Medications Department (see advertisements). They ship out the order next day so there is no long delay in getting the product. This is the vitamin that we have used for many years and highly recommend it.

Do not overfeed your birds. Overfeeding and lack of exercise does many of our birds harm. Other than inbreeding, obesity is a major cause of egg infertility in Chinese Painted Quail. If you have the time, the best plan is to regulate the feed intake carefully. This means that time will have to be taken to feed the birds a small amount in the morning and in the evening. Since Chinese Painted Quail are high energy burners, they need feed every so often during the day. The amount of feed intake and the feeding schedule can be decided by seeing that the birds are always just a little hungry. The time of year, the condition of the birds, and other factors will cause feed intake to vary.

Nevertheless, feed intake is also a good way to gauge the health of the birds. When they go off their feed, there is always a reason. It could be the climate or another reason that would be of no concern. But, it could mean "trouble in river city" indicating that there is a disease, a parasite, or stress present. Careful observation of the birds will tell.

We regularly **feed mealworms** to the birds. Chinese Painted Quail of all ages love them. We constantly have a crop of mealworms available for our birds. Mealworms can be grown by the breeder.

CHAPTER 5

HOW TO RAISE CHINESE PAINTED QUAIL

One of the most prevalent problems in breeding Chinese Painted Quail is inbreeding. We probably know better, but we still go ahead and do it. We get a pair of birds and have some success. Then, the young are sold to other breeders and they mate them together (brother and sister). The next breeder does the same thing and on and on. In only a few generations we have worthless stock.

Do not inbreed. Do not breed brother and sister or any other close relative together. If this is done, the offspring will become sub-standard. Signs of inbreeding are small size, infertility, general listlessness and ill health. Get yourself a second pair of Chinese Painted Quail from another source so you will not inbreed.

Natural Brooding

Letting the birds raise and brood their young themselves is one of the most satisfying parts of aviculture. In the early days, this was the only method of raising birds in captivity. It is certainly not the most efficient. Those that do this report that many more of the young are lost than when the young are raised artificially.

We let a pair of quail raise their own chicks one year. I must say, it was a very rewarding and interesting experience. We saw the parent birds in an entirely different role. Their personalities changed to meet the role that they played in raising the chicks. They were successful and proved to be very good parents.

If you have only a pair or two of Chinese Painted Quail, you may want to try to let them raise their young naturally. Given the right circumstances they will do so successfully.

> HINT: Make the pen as natural looking as possible with plenty of privacy.

Put them in an aviary which has a dirt floor. This will enhance your chances of success. They just do better in a ground pen. Put in some natural clumps of tall grass in one corner and arrange a log or two so they will feel hidden. It is a good idea to put sharp river sand in one corner. Be sure that it is an area that is dry. The cock bird will guard his corner assiduously and will keep all intruders away from "his bedroom." The couple will do much better if there are no ground birds in the pen. The loving pair will do quite well if there are other birds in the aviary as long as they are the type that live in the upper pen. Be sure that the wire is small enough around the bottom of the pen to keep the tiny chicks from wondering. I have had them escape from the quarter inch mesh so it is a good idea to use the one-eighth inch wire. It is amazing how chicks can find just a tiny hole and get out and away from the care of the parents. If the weather is cold they may chill and be lost.

Place the birds in the pen about two months before their normal breeding season. Early Spring is a good time to put them in their new home. This will give them a chance to get adjusted before they come into breeding condition.

Give them high protein feed as they come into breeding condition. Mealworms are an excellent source of natural protein.

Disturb them as little as possible during this time. When you see the hen missing during the day you can be sure that they are planning on a family. When you see their droppings change from normal and become large with more white pigment than usual, you can be sure that they are setting on eggs. If you can contain your curiosity, do not try to find the nest in a heavy planted aviary.

The hen is the only one that usually sets on the eggs. The cock bird will guard the area while his hen incubates the eggs. If the cock does not leave the hen alone while she is incubating, it is best to remove him. In 16 days the eggs will hatch and the little mother will bring her brood of eight to ten out to see the world. When the chicks hatch and come out, watch the cock carefully as he might peck at them and cause injury. More than likely, he will adjust to his newly acquired fatherhood and do just fine. You may even see him calling the chicks to feed them a treat.

> HINT: The eggs may hatch earlier than you think if the hen begins to incubate before you know.

If the pair is to have success, they must be in moderate temperature. If it gets too cool at night the chicks may chill early in the morning as they scamper around finding food. Plan ahead and put the pair in a well suited aviary before they lay.

Feed the chicks Turkey Starter along with small seeds. Crumbled bread is appreciated. The adults will see that the chicks eat if the food is available. Offering a variety of foods will insure a balanced diet.

If all goes well, the chicks can be judged on their own at about four weeks. They should be removed from the area as the cock will again come into breeding condition and shield his domain from his sons.

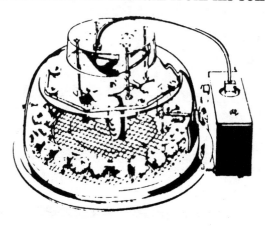

This little Marsh Farms Turn-X incubator can be used for incubation, hatching, and then for brooding a small number of quail chicks.

Artificial Incubation and Brooding

There is no doubt that this is the most efficient way to raise Chinese Painted Quail. Some of the captive bred and raised birds just will not incubate and raise their own young. Regardless of the reason, more chicks will be hatched and raised to maturity when we assist Mother Nature by hatching the eggs and raising the chicks artificially ourselves.

If we plan to do the job of the hen, we must have some type of incubator. There are basically two types of incubators. We have used both with success.

> HINT: Some styrofoam is not fire retardant. Be sure and have a thermostat that shuts off the heating unit at 110 degrees.

The Still-Air Incubator. This is the cheapest type of incubator. This unit has some type of heating element which is capable of keeping the eggs at the proper temperature during development. Some commercially manufactured still-air incubators are made from styrofoam. This material serves as a very good insulation but some is not fire resistant so great care should be used.

There are several of these miniature units on the market that would be perfect for the person that wants to hatch a few Chinese Painted Quail eggs. They are as inexpensive as $35.00. I noticed a "do it yourself" incubator kit that sells for about $25.00 which has the heating element provided along with plans to build the box.

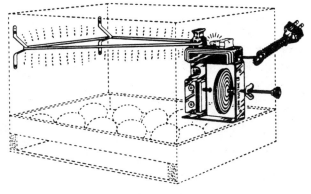

Supply houses have incubator kits available for the do-it-yourself breeder. This illustration shows chicken eggs but it will work fine for quail eggs.

We have hatched hundreds of quail in a home made still-air incubator. This is made from 3/4 inch plywood and has a piece of thick 3/16 inch glass as the top. It measures about 14 inches square. The heat comes from a special wire heating element that is in the top near the

87

glass. The water pan for humidity control sets on the bottom below a wire false bottom on which the eggs rest. We hatched our first Chinese Painted Quail in this little incubator. We still use it today many years later. One of the drawbacks to this unit is that the eggs have to be turned by hand at least three times a day. They are simply rolled over the wire to a different position. Some of the more expensive manufactured incubators have automatic egg turners.

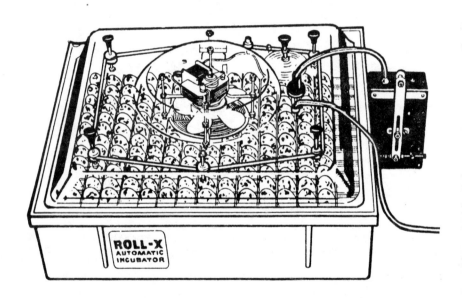

This Marsh Farms Roll-X incubator can be ordered with a special turning grid for Chinese Painted Quail.

Any good incubator must have a dependable **thermostat**. It should be accurate enough to keep the eggs at the proper temperature at all times.

In still-air incubators, the temperature reading should be made at the top of the egg. By measuring at this level, the temperature at the center of the egg should be the ideal 99 3/4 degrees. The correct temperature at the top of the eggs should be about 101-102 degrees which will give an average 99 3/4. To find out what the correct reading for your eggs should be, take a reading at the **bottom of the eggs** and then take a reading at the **top of the eggs**. If, for example the temperature at the top of the egg is 102 degrees, and the temperature at the bottom of the egg is 97.5, the average temperature (add top and bottom temperatures and divide by 2) is 99.7 which is close to the ideal of 99.75 degrees in the center of the egg. Your incubator will need to be adjusted to get the ideal average temperature.

> HINT: These still air units make excellent hospital units when not in use for hatching eggs.

The Forced Air Incubator. This is perhaps the best type of incubator. Of course there is a price tag on them. They are much more complicated than the still-air type as they have a fan causing a steady flow of air to circulate over the eggs at all times. Some of the better ones have electronically controlled thermostats which are very accurate.

We have several of these and find them very convenient and efficient. The only problem is that most of them are probably too big for the small breeder that wants to hatch only a few chicks each year. There are some very good small still-air incubators produced for the small breeder in mind. Always follow instructions.

There is one on the market that can hatch the eggs and then be converted to a handy brooder for the young chicks. (See advertisements)

HINT: A complete discussion on incubators and egg hatching can be found in the author's book *RAISING GAME BIRDS* (see page 31-44)

Often, we are asked which type of incubator we prefer as we have several of the different name brands and even home-made incubators. There is no incubator that will fit everyone. Consider the cost, purpose, type of birds, and parts availability when buying your incubator. We use and recommend the Marsh incubator units (see advertisement).

The Assembly Line Method

When the chicks hatch, the breeder must provide the comforts that small chicks require to grow and mature. For years it was thought that these tiny chicks were far too small to be raised artificially. Modern procedures have proven this thinking to be false.

We have a method which takes ideas from others and puts them with our own experience. Over the past forty years we have come up with a method of raising chicks that works for us. We call this method the **Assembly Line Method.** It is not a new technique, nor is it one that works only with one particular species. It can easily be adapted to raise the chicks of any game bird that is kept in captivity. The George Allen family at the Game Bird Preservation Center in Salt Lake City use the same principles to successfully raise hundreds of waterfowl each year.

The **Assembly Line Method** entails the use of different "surroundings" (boxes) in which the chicks mature through each of the growth stages of their lives.

The needs of the chicks change as they mature, so this method changes with them. The chicks represent an automobile on an assembly line which places separate parts on the chassis at the correct time. The needed part is available at the needed time.

This basic method of raising chicks can be used in any climatic conditions. Many waterfowl and pheasant breeders use this same basic method with success. The only difference is the dimensions and sizes of the feeders and waterers. The boxes are mobile and can be placed in a warm room if the outside temperature gets too cold.

> HINT: It is best not to have mature birds in the same area with chicks.

Each hatch (usually 4 to 20 chicks) is placed in a box which is modified to meet their basic needs at a particular age. If more than 20 chicks are hatched you may want to break them into two groups. This is especially necessary if the chicks in a particular brood are prone toward cannibalism. We have a series of boxes which house the different age chicks.

The Nursery Box

This is the first box (environment) on the **Assembly Line**. It is used to house the chicks as soon as they leave the incubator and serves as home for the first two weeks. The Nursery Box is about 15 by 20 inches in size and about 20 inches high. This box should be high enough so the chicks cannot easily jump out. It can be an ordinary cardboard box or can be custom made from wood. Be sure to get a box that is longer than it is wide so the chicks can have a place to get out of the heat if they want to do so.

The heat source is a light bulb of 40-watts placed about two inches from the floor, but above the heads of the chicks. If this is not enough heat for the young chicks you may use a 60-watt bulb, but care should be taken to insure that it does not get too hot. Also be sure that the light bulb does not hang too close to the sides of the box as this presents a possible fire hazard. We spray paint these light bulbs red which seems to stop "glare blindness" and reduces picking. Common spray paint is used making sure it is the fast drying type. Painting the light bulbs seem to shorten their life, but this is a small price to pay for sound chicks. If "infra-red" heat lamps are used, great care should be taken to prevent over-heating the box. Most of these types of lamps come in

93

150 or 250 Watts which is too much heat for a small box. Watch the chicks to see if they are comfortable. They will tell you if they are too hot or too cold.

> HINT: Many breeders use two small wattage light bulbs in case one should burn out during the night.

The Nursery Box has a cloth floor made out of an old towel that can be washed and reused. The towel offers good footing so leg problems are thwarted. This towel bottom should be changed every day. The more chicks in the box, the more often the towel, feed and water should be changed. Do not let the towel become wet or matted with droppings. Be careful to see that no loose strings are on the towel as they will be eaten and can strangle the chicks. A real danger is to get the chicks too crowded. More than 20 chicks in one of these boxes causes problems. If you want to put more chicks than this together, the box should be greatly enlarged. Some species of game birds do much better if more than two or three chicks are raised together. There seems to be something to their liking if the youngsters are grouped in numbers that are natural to them in the wilds which is usually from 5-7.

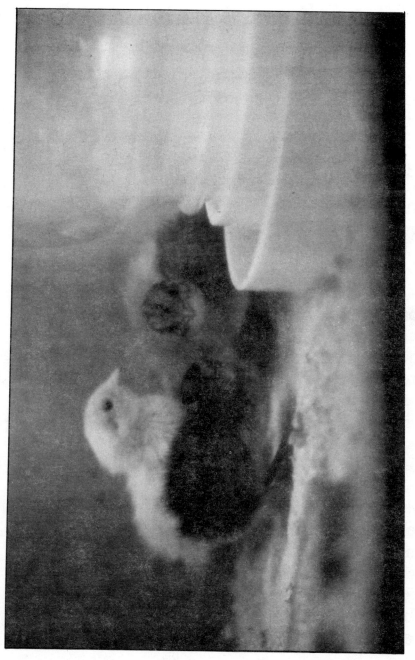

The color of the chick down tells us what mutation the chicks are when they hatch.

Starter feed is scattered around on the towel. We use starter feed with 28% to 30% protein. This can be turkey starter or game bird starter that is made by any good feed company. The turkey starter has a medication which does not hurt Chinese Painted Quail at all and helps control Coccidiosis and Blackhead. Usually this feed is too course for the small chicks so it should be crushed into very fine crumbles. My wife uses her food processor or blender to do the job quickly. Feed which will last several weeks can be crushed at one time.

The water fount is placed out from under the heat source so the water will not get hot. Warm water breeds bacteria and can be harmful. The water kept out in the cool area also gets the chicks out running. They need exercise to build healthy bodies. There should be several water founts if you have many chicks. Two or even three are advised. Do not worry about the feed on the floor as it will not hurt the chicks if they eat off the floor for a few days. It is essential to have the feed accessible so they will learn to eat quickly and get a good start. Older chicks should never be fed off the floor. It is a help in getting the chicks to eat if they can pick at things on the floor. Diverse colors seem to attract their curiosity. Sprinkle some chopped lettuce on the floor. If they pick at it and it tastes good, half the battle is won. They will get the taste of the chick crumbles and will learn to eat rapidly.

Rocks or marbles can be placed in the founts so the babies will not get wet and chill or even drown. The regular poultry water founts are too tall for the baby quail, so jar lids can be used if care is used in placing the rocks in the lid. When small quail chicks are started it is much better to buy specially molded plastic water bases. These fit on everyday canning jars and can be bought from suppliers. Some type of medication should be put in the water the first two weeks. Bacitracin, Terramycin, or some other antibiotic works fine.

HINT: Very cold water should never be given as under certain circumstances the young chicks will die after drinking chilled water.

The top of the box should be covered with hardware cloth (welded wire) to keep the active chicks from jumping out. We put newspaper over a part of the top to keep the heat in the box. This newspaper can be adjusted to control the heat inside the box by leaving more or less space open. Again caution should be used to keep the bulb from being a fire danger.

We have found many times, more chicks are lost due to overheating than by not having enough heat. If the chicks bunch together and chirp they are cold and should have more heat. If they scatter and sleep away from the heat, they have enough heat or maybe too much. Panting chicks also indicate too much heat.

The Baby Box

After the chicks are about 10 days old, they can be graduated to the second box in the **Assembly Line.** This box is about the same size as the "nursery box" or it can even be larger. It has a special hardware cloth (welded wire) bottom that rests about 2 inches or so from the bottom of the box. It is designed this way to let the droppings fall below. It should be the aim of the breeder to keep the birds away from their own droppings as much as possible from this age through maturity as young birds are very susceptible to many diseases spread through droppings.

The welded wire should be the small size, (1/8 inch) as the small feet of the babies will go through the larger 1/4 inch wire.

The light bulb should also be adjusted and the wattage changed to less heat output. If too much heat is given to the chicks at this age they will not feather out

> HINT: The heat source is very important in keeping the chicks comfortable. Be sure to have the heat well above the backs of the chicks to avoid "bare backs."

properly. The rocks or marbles can be omitted from the water founts. Feeders should be given to the chicks. We use flat lids with some kind of wire guard to keep the birds from scratching out the feed.

By this time the two-week old chicks will be feathered enough to fly, so care should be given to prevent them from flying out when the wire top is removed for servicing the chicks. Each time we move the chicks, their wings are carefully clipped. The primary wing flight feathers are clipped on only one wing since the tops of the pens are not very high.

If the pens are very high, clip both wings so the bird will stay balanced when it tries to fly up. If the are unbalanced, they will likely break wings or legs as they crash. Wing clipping in our experience is very necessary to keep the chicks from escaping and scalping the tops of their heads.

The Juvenile Box

This is the third box which is the same as the second. It can be larger if desired. To avoid crowding, we have several of these juvenile boxes so the chicks can be divided into smaller groups to prevent crowding. We have found that one of the most common problems in growing quail is overcrowding. This is very dangerous as cannibalism, disease and other problems may occur under these conditions.

The black eyes show that this is not an albino. This white mutation is hard to sex as both male and female are alike except for the larger size of the mature female.

100

Inside view of plywood growing pen showing dust box, feeders, and waterers.

HINT: A dust (sand) box is imperative if the feathers of the birds are to be kept in good shape. The appearance of your birds will greatly improve if they have a "dust bath" at least twice a week.

Clean sand is good to dust in. Do not be alarmed if the birds eat some of it to get their grit.

101

Mixing assorted ages of chicks can be done if a few rules are understood. As a general rule it is best not to mix species. Never under any circumstances mix Bobwhite Quail chicks with other species.

Mixing different ages of Chinese Painted Quail can be successful if you do it when you move the birds from one box to another. It seems the different box is new to everyone and mixing can be done. For this reason we arrange each of the boxes in a different way so it will not look like the other boxes. By the time the chicks get used to the new box they are used to the new neighbors and all get along well.

Never, never add a new chick or group to a box already occupied. The residents will take offense to the newcomers. We learned this lesson the hard way. One of our favorite chicks that was the only one to hatch in that particular batch of eggs was lonely so we put him in with some other chicks. To our dismay he was picked on but we were able to rescue him in time to save him.

This is an area where the breeder should use good old common sense. Think smarter than the birds. Use your superior intellect to outsmart them by taking advantage of their habits. There is no perfect method of raising birds. Yours will work for you.

HINT: When adding or forming new groups of chicks, always change the inside arrangement of the box. Do not add any new chicks to a box already occupied.

The Growing Pen

The fourth area in the **Assembly Line** is put into use when the birds have feathered out enough to be without heat at night. It is much larger than the "boxes" in order to give the birds more room. It can be a mobile pen that is moved from place to place, or it can be a permanent ground pen built outside.

The Mobile Growing Pen. We make these pens out of 1/4 or 1/2 inch plywood. The size can be varied to meet your needs. This simple pen is constructed as follows:

First, cut three (3) pieces off of the 4x8 sheet of plywood which measure 2x4 feet. Cut one of these in half giving two pieces 2x2 feet which will be used for the ends of the pen. You should have two (2) pieces 2x4 and two (2) pieces 2x2. Next, cut out windows in each of the long pieces of plywood roughly six inches high by three feet long. Make these long windows about two inches from the bottom edge of the side pieces. Cover these areas with small wire mesh to give light during the day. Cut a door in one end (or both ends), hinge it and put a latch on it (or have a sliding door system). Nail the pen together using 2x2's in the corners. Place strips of this 2x2 stock along the bottom edges so there will be a nailing place for the wire bottom. Cover the bottom with wire hardware cloth (welded wire mesh) 1/8 inch in size. Cover the top with 1/4 inch wire. This will enable you to simply turn the pen upside down to have a different size wire bottom to favor the size of the quail's feet. You can get either mesh size hardware cloth in two foot widths. A piece four feet long will fit the pen area just right.

You should have a box-like pen with a wire bottom and a wire top assembled. If you decide to stack these pens on top of one another to save floor space, there needs to be a dropping board put on top of each pen. We use newspapers on this board, so keeping it clean is easy. If the pen is not placed out in the open where it can get wet from rain, we use heavy cardboard tops.

A night light encourages the birds to eat and drink during the night, and also makes them calmer.

> HINT: Place a light in one end of the pen above the feed and water area. This will not be used for heat, but will be an attraction light for the birds at night.

The birds can stay in a pen of this type until they are mature if not too crowded. A pen like this can handle up to 40 Chinese Painted Quail if care is taken to prevent picking. We put alfalfa hay in the bottoms and keep the beaks seared off.

The Outside Growing Pen. If the young birds are completely feathered and able to stand the climate without any heat they are placed in outside pens. We clip both wing primaries at this time to prevent "unbalanced" flying which can cause injury. If your climate is wet you need to make some of the pens described above which can be kept indoors. Do not put the birds on the ground unless the climate is dry. The measurements of this outside ground pen can be about anything the breeder is content with. Ours are four (4) feet wide by eight (8) feet long. The idea is to make the pens longer than wide to give the birds plenty of room to run back and forth for exercise and probably just as important, give them a place to retreat when they are threatened.

Have a haven in one end of this pen where feed and water is kept out of the weather. We use platforms made out of 2x2's which are covered with wire mesh to keep the feed and water off the ground. You may find that the birds refuse to get in the shelter and prefer to sleep out in the open. If this is the case give them some branches out in the open. It is a good idea to have 4-5 inches of sharp river sand in the bottom of the pen. If the grass has grown up in this ground pen in early Spring it may be necessary to mow a strip or two down the sides to give some running room. We like to keep the ground bare with no vegetation at all. If the ground is wet enough to sustain vegetation, it also supports disease agents, earthworms, and other creatures.

A good general rule is to make the birds feel as snug as possible. Also, give them as much security as feasible so they will breed better. They begin breeding at six weeks of age. Our pens are made as natural as imaginable with brush, logs, and clumps of tall grass which is pulled up by the roots and put into the pens. The tops of this grass usually falls over providing hiding places.

Be careful not to keep mature males together. They mature very fast so be watchful. If you must mix several males together put them in a pen out of sight of any females. It is best to have them out of sound so they cannot hear the calls of the females which stimulates fighting. The males may have some small skirmishes until the pecking order is established. After this is done, they will settle down if no hens are around.

There are some exceptions to the above. Males will get along in a colony situation. (Large pen with several pairs of birds in it.) If males have been raised together they seem to get along better than if total strangers. One breeder recommends that at least 3 males be put into a colony breeding situation.

There is no hard rule here. This is one of those situations where it is best to assume the worst which means the males will fight. If they do not fight you have a tolerable arrangement. However, be on guard as birds like people change their minds or have squabbles which could lead to serious fighting.

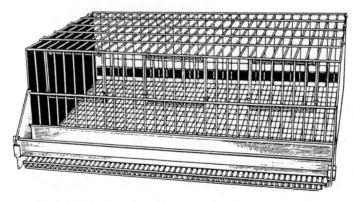

These specially made units allow the eggs to roll out on the porch in front of the door.

107

Some breeders have set up an automatic watering system which requires very low maintenance compared to individual waterers. The automatic system must be cleaned out periodically to keep algae from forming during hot weather. This can easily be done by flushing out the system with disinfectant that kills algae and bacteria.

A low concentrate of "bleach" can be added to the water to keep it clear and clean. One tablespoon per five gallons of water is a good dilution. This will work only with systems that uses a gravity flow container. This will not work if your hose is attached to the water supply.

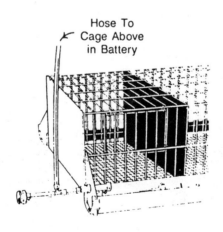

Battery breeder unit showing details of an automatic watering system.

CHAPTER 6

KEEPING CHINESE PAINTED QUAIL HEALTHY

Disease and Symptoms

The number of diseases that affect Chinese Painted Quail are about the same as for any upland gamebird. The chances of the average Chinese Painted Quail getting any serious disease is very remote. There are some common (and minor) disease problems that at one time or another every game bird breeder will encounter. It is the purpose of this chapter to discuss these diseases in jargon that is easily understood by the non-medical person.

I remember, many years ago, when I read my first poultry disease book! I was scared so much that it was some time before I got enough nerve to get any kind of birds at all. The descriptions of the diseases were done in such a way that I just knew that my birds (when I got some) would get every one of those diseases described in the book. It turned out that the problem of disease was over-stated in that case. I did loose my quail to Coccidiosis before the modern medicines were available.

When a disease problem does break out it is **no disgrace or indictment** to anyone. Only with the few perilous and highly contagious diseases should there be any real problem. In all cases where mortality (death) is high, it is recommended that the breeder seek professional help. A veterinarian trained in avian medicine and the use of a good qualified lab is a must in diagnosing many disease problems. Sometimes it is hard to find a qualified vet in the local area. We suggest that you use your State University Veterinarian School or your County Agent to get their recommendations and referrals.

A Symptom is not necessarily a disease. I once knew a man who became very upset every time his birds got a little off their feed or did not look "just right." I compliment him on being observant enough to see the subtle changes that his birds went through. However, every time a bird shows what we would consider a deviation does not mean it has a disease. Birds have good and bad days just like us.

110

> HINT: Birds sometimes look droopy for no reason. If your birds look droopy for more than a day or two, check to see why.

Symptoms are signs that show us there is a disorder of some kind. It may be serious or it may not be serious. The diagnosing of a disease from symptoms should be left to the expert veterinarian. However, there are some things the breeder can do before he calls for help. The problem is that some diseases have many symptoms that are common with many different diseases. "Which way does the road lead when there are so many different signposts?"

The wise bird breeder will note all of the symptoms and if they do not go away in a day or two and if the birds begin to die he knows that there is a serious problem. When to seek professional help can be decided by the seriousness of the symptoms and the mortality. Sometimes, it takes a disease many days to bring death to its host. Other times the bird can die in just a few hours from the first sign of any symptom.

Occasionally you will have a bird die. This can be ignored if it does not continue. Anytime there are more than two birds dying from some disease, you should consider the problem to be **contagious** and should seek help immediately. The use of isolation cages will help keep the problem within a limited area while you solve your problem.

How Disease Spreads

There are some things that can be done to keep Chinese Painted Quail vigorous and disease free. An understanding of how disease can be spread gives us many ideas. Probably this is an area where good hygiene or sanitation does more than anything else the breeder can do. You do not need to be "hospital sterile" but be as clean as feasible.

When it comes to disease problems a good general rule to follow is to use everything that is available to discover and treat the problem. Typical disease is spread by one or a combination of many of the following ways:

❑ **Ground and pens that are contaminated.** It is common sense to keep the bird area clean.

❑ **Airborne organisms.** Disease agents are everywhere and blow with the wind. Keeping dust down will help, but there is nothing that can be done to completely do away with airborne organisms.

❑ **Rodents and wild birds.** These pests can not only carry diseases, but can be costly if they rob the feed pan.

❑ **Newly acquired diseased stock.** A good practice is to put new stock in a quarantine area for several days before putting with resident birds.

❑ **Egg transmission.** Many bird breeders get hatching eggs. When getting hatching eggs, fumigate the eggs before they are put into the incubator.

HINT: There are several very good books on diseases of birds that may be valuable. These go into great detail and if studied carefully will give more information about this important subject. Contact your state veterinarian school for a list of books.

A Quail Miracle Drug

When I started raising quail as a boy (forty years ago) there were no drugs available for birds. Now things have changed! We have been using a drug called "Bacitracin". This is the generic name. The product name of the drug that we use is "Solu-tracin".

The product is especially made for birds and it really works. It is an aid in the prevention and control of Enteritis which is probably the worst quail killer. For several years we were distributors of this product and sold many packets to quail breeders. This is the only way that I know of to keep quail alive when kept on the ground. The dosage that we use was worked out with the help of a veterinarian up in Oregon several years ago. Using this dosage we raised over 500 Mountain Quail on the ground and many hundreds of other species. The dosage method is as follows:

As a prevention: one teaspoonful per gallon.

As a treatment: two/three teaspoonfuls per gallon.

Give as the only water for a minimum of 7 consecutive days and as long as 10 consecutive days. At first the birds will hesitate to drink as it is rather bitter,

114

but soon they will drink with no problems. In order to get the drug into the bird's system it must be given the recommended length of time. Bacitracin must not be given on a continual basis as the gut will be sterilized of all bacteria (including the good guys) and the birds will get fungus and other infections usually warded off naturally.

For susceptible birds give the above dosage once a month skipping at least 20 days between dosages.

We give the drug in the prevention dosage to newly hatched chicks to help them get off to a good start. It works for us!

Keep A Medicine Chest

There are some basic medical items that every bird breeder should have handy. One never knows when any one item will be needed so it is best to have them handy at all times. A good source for most of these items is the local feed store and your local druggist. There are some mail order houses that deal specifically with these products.

Some of these are as follows:

- ❑ Basic first aid materials (scissors, tape, gauze.)
- ❑ Various insecticides (liquid and dust.)
- ❑ Broad spectrum antibiotics. (Aureomycin, Bacitracin, Terramycin.)
- ❑ Worm medications. (Tramisol, Piperazine)
- ❑ General disinfectant.
- ❑ A sanitary container for the above items.

A supply house that specializes in medical equipment will be a good source for these items. One of the larger chain drug stores will have many that can be picked up separately or in a first aid kit. Your druggist will help you in getting together the medications. Your general farm store will have some of the medications you need.

> HINT: Have a special place to keep your medicine chest. It should be handy to the bird area.

Using A Hospital Cage

Some kind of "hospital cage" should be kept on hand to use when a bird is not feeling well. Many times there is no disease involved when a bird gets out of sorts. They need some rest or relief from stress. Old-timer bird people know the value of applying heat to a sick bird. The bird's system many times can take over and give relief enough for natural healing.

A good hospital cage or box can be made by any handyman. All it needs is a good reliable source of heat, no drafts, good sanitation, adequate humidity and available food and water.

Some very good cages or boxes can be bought commercially with thermostatically controlled heat. These work very well but can be expensive. We have found that a simple cage or box works just as well as these costly ones.

Hints From Hayes

Many game bird breeders like to have advice or hints to go by which help them be a better breeder. An engrossing thing about raising quail is that many times there is no *right or wrong* way to do it. Whatever works for you is right. My methods have been certified successful by myself and many others. However, my way is not the only way.

Listed below are some *Hints from Hayes* that will give any astute game bird breeder some spring boards to use to develop a system of keeping their birds healthy. Some like to keep a check list to insure that nothing is forgotten.

❑ Always get good quality merchandise.
❑ Build pens for welfare of birds.
❑ Buy from reliable breeders.
❑ Feed your birds going from young to old.
❑ Clean between groups of birds.
❑ Clean and disinfect waterers daily.
❑ Clip wings when necessary.
❑ De-beak when needed.
❑ Develop a simple system.
❑ Do not weaken stock by inbreeding.
❑ Eggs brought in have least disease.
❑ Experiment with new methods.
❑ Feed treats.
❑ Feed the best feed available.

- ❑ Fumigate incubators regularly.
- ❑ Get early diagnostic report.
- ❑ Isolate immediately any sick bird.
- ❑ Isolate after bird shows for 30 days.
- ❑ Keep old birds away from incubators.
- ❑ Keep pens and water clean.
- ❑ Keep toenails clipped.
- ❑ Keep species separate.
- ❑ Maintain good records.
- ❑ Medicate when needed.
- ❑ Never let birds get to their droppings.
- ❑ Never mix different age birds.
- ❑ Never permit water to puddle.
- ❑ Proper management is essential.
- ❑ Properly care for hatching eggs.
- ❑ Raise the birds you like.
- ❑ Safely destroy dead birds immediately.
- ❑ Sell only healthy birds.
- ❑ Use a hospital cage.
- ❑ Use safe water.
- ❑ Use vitamins.
- ❑ Use good, common sense
- ❑ Worm your birds.

Hints On Ground Pens

If Chinese Painted Quail are kept on the ground for any length of time the breeder must take certain actions to insure that the ground not be contaminated by unwanted organisms. There are several parasites and forms of bacteria that use the ground as breeding places especially if the soil is wet or damp. Dry soil discourages organism growth. Dampness must be eliminated if the birds are not to be infected by parasites or disease.

To the bird breeder, the normally beneficial *earthworm* can be a real problem. The reason the earthworm is harmful is that they play host to some detrimental organisms which live in and under the soil. The disease called Blackhead uses the earthworm as host in its life-cycle.

A very good method of ground disinfectant that we have used successfully is to treat the soil at least twice yearly using the following method which treats about 100 square feet of ground. We first learned about this from an article in the *GAZETTE* (Allen, 1967.)

Two common chemicals are needed. First, eight pounds of "slacked lime" ($Ca(OH)_2$) which is readily available from any builder's supply house. It comes in 100 pound sacks and is a very fine white powder. Face masks should be used to avoid breathing the material.

The second chemical is four pounds of powdered sulfur which is available from most agriculture supply houses. The common "soil sulfur" will work just fine and is less expensive than the sulfur used to dust for insects.

> HINT: The lime and sulfur treatment lasts for about six months. It helps control mites as well as keeping soil clear of bacteria and other invertebrates.

Mix the two chemicals together thoroughly and spread as evenly as possible over the top of the ground to a depth of about 1/2 inch. Turn the soil over using a spade. Take a rake and smooth the ground. Then take some more of the mixed chemicals and spread it over the ground evenly to a depth of about 1/4 inch. The pen is now ready to be occupied. If the birds are tame, you can treat the soil without removing them. We think it best to remove all feed and water and give the pen a good cleaning before treating the soil.

If the topsoil is removed at least once a year (preferably twice) the above treatment will not be necessary. Many successful breeders remove about 5 to 6 inches of soil each cleaning time and replace it with river sand. This is probably the best thing to do if possible. Remember to keep the pen area as dry as possible to keep bacteria growth down and also to keep the other creatures from using your quail pen as a breeding ground. This is controversial among breeders. There are many that want the aviary planted to give it the natural look. This is a beautiful thing to see, but the well being of the birds must be considered. I have to say that I have seen pens both ways that were successful.

A Good General Remedy

Dr. Tom Smith, Extension Poultry Specialist with Mississippi Cooperative Extension Service, gives us an excellent general remedy which will help keep our birds healthy. It is a solution of common aspirin which is used as a general treatment for reducing distress conditions of birds (fever or listlessness) that accompanies many diseases symptoms.

Dissolve five (5 grain) aspirin tablets in one gallon of water. Offer this solution free choice to the birds for the duration of an illness.

CHAPTER 7

SOME COMMON DISEASES

It is hoped that this section will be useful in the recognition, treatment, and prevention of some of the diseases which infect our birds. We have not included **every avian disease**, but have chosen to include some of the disease problems that will most likely be encountered in every part of the world.

There are many other disease problems that can affect these little quail. The reader is referred to the author's book *Raising Game Birds* for detailed discussions of avian diseases. (See advertisements) There are other good books available that will give guidance to the reader that wants more detailed and unabridged facts about avian disorders. Your book store can order some of the more popular books for you. If you cannot find any titles to buy, then check with your local lending library.

The following mixtures are recommended to flush the digestive system of toxic substances. These solutions will work anytime the system of the bird needs to be flushed because of poisons or even general constipation problems. It is a good idea to watch your birds after treatment to determine if the dosage could be reduced as it is easy to overdose a tiny bird like a Chinese Painted Quail.

Molasses Solution: Add one pint of molasses to five gallons of water (or 1/5 pint molasses to one gallon). Offer the drinking solution free-choice to the affected birds for about four hours. Treat severely affected birds individually if they cannot drink.

Epsom Salt Solution: 1/5 pound of Epsom Salt per one gallon water for one day. If the birds are unable to drink mix one teaspoon of Epsom Salt in one fluid ounce water. Place the solution in the crop of the bird. This amount of solution will treat about eight quail.

Kerosene Drench: Used in conjunction with one of the above laxative treatments. Two tablespoonfuls of solution (one tablespoon of kerosene in one cup of water). Give orally.

QUAIL DISEASE
(Ulcerative Enteritis)

This problem is often called **"quail disease"** which until controlled and understood was perilous. It is probably the most typical "killers" among quail.

When *Clostridium colinum* builds up the disease occurs.

SYMPTOMS: All age quail get this disease. Birds can die with Enteritis while being in good flesh and have little or no symptoms or lesions. Other birds will appear listless, ruffled feathers, white diarrhea, and have a "humped-up image."

To verify the disease a lab should be used.

Many times the birds have **Coccidiosis and Enteritis simultaneously** and thus show blood in the dropping.

TREATMENT: The newer drugs seem to be successful in the treatment of this disease. Bacitracin or newer drugs may be used.

COCCIDIOSIS

This disease is a very common one found in captive-raised birds. It is probably the first problem a new aviculturist will encounter. Thankfully it is not now serious. It wiped out whole flocks not more than 30 years ago. Birds raised on the ground are very susceptible while birds raised on wire rarely get the disease. The causative agent, a *protozoa* is found in several forms. The best control is to break into the life cycle with drugs and good sanitation thus avoiding multiplying of the organisms. Young birds are more susceptible than adults. Recovered birds have a resistance to the particular strain they recovered from, but can get the other strains. There are a number of strains that affect Chinese Painted Quail.

SYMPTOMS: Weakness, ruffled feathers, hunched posture, and droppings that may be bloody are signs of this disease. Infected birds are inactive and become less interested in feed and water as the disease progresses.

DIAGNOSIS: The presence of oocyst will confirm the disease.

PREVENTION: Coccidiosis can best be controlled by prevention or letting the birds get a mild form thus building up immunity to the disease. Keep the area sanitary and avoid the ingestion of droppings if possible. The recognized ways of control are:

- ❑ Mild case to build immunity.
- ❑ Feed coccidiostat from first day to 14 weeks.
- ❑ Then treat outbreaks.
- ❑ Feed high levels of Vitamins A and K.
- ❑ Treat any outbreaks as they occur.

> HINT: This is not a complete discussion of these disease problems. Those interested should check with their local lending library for other titles.

TREATMENT: There are several good medications available for this disease. It is a good idea to have medication on hand for treatment and control.

CHAPTER 8

SOME COMMON PARASITES

All Chinese Painted Quail will have some sort of parasite during its lifetime. It is a sure thing to assume. Years ago, parasites were much more of a threat than they are today. Modern knowledge of the life cycles, modern medications, and modern preventive measures has made the parasitic problem a minor one if the breeder eradicates on a regular basis.

Generally speaking, there are two types of parasites. (A parasite can be described as an organism that is foreign to its host and which has the potentially of doing harm or even causing death).

I will discuss some of the parasites that are likely to be a problem with your Chinese Painted Quail. You probably will never encounter most of these if you keep your birds clean and healthy. However, I would certainly recommend that you worm and dust your birds with some sort of insecticide powder at least two times a year.

You may want to develope the habit of dusting the birds every time you catch them up for whatever reason. There are several good products that contain the chemical **"Sevin"** which can be bought at your local Farm and Feed Store.

LICE

These insects are the most common and widespread external parasites of Chinese Painted Quail. Lousiness of birds can be diagnosed by finding on the birds wingless, flattened, brownish-yellow, quick moving insects. Lice spend their entire life cycle on the body of the bird. Eggs are attached, often in clusters, to the feathers.

The entire life cycle takes about two or three weeks for completion. One pair of lice may create 120,000 descendants within a period of a few months. Their normal life span is several months, but away from the birds they can remain alive only five or six days.

The most common way for the spreading of these parasites is through body contact. At night they are active and probably crawl onto the roosting quail. If you find only one louse you can be sure that they have infected your entire flock of quail. In this case begin a systematic effort to break their life cycles before they become epidemic.

TREATMENT: Lice can be controlled by regular dusting of the birds and house areas with an insecticide powder. Sometimes it takes several applications before the vermin are controlled as young are constantly hatching out until the life cycle is interrupted.

MITES

Most mites that attack Chinese Painted Quail use blood for food, therefore anemia is a more or less common symptom. As might be expected, mites can easily transmit many bacterial and viral infections because they are bloodsucking insects. Mite infections seriously lower disease resistance and they may actually kill the birds through the extraction of blood. It is best to check the birds at night. Do this with great care as Chinese

Painted Quail are very sensitive during the night. It is probably best not to disturb the birds unless absolutely necessary.

Some of the different mites one may encounter are Red Mites, Northern Round Mites, Chigger Mites, and Scaly Leg Mites. The climate and geographical location will determine which mite is indigenous to your particular area. Chiggers are a common problem in the South where the climate is warm.

Breeders should be aware for outbreaks of Mites. This is especially true in a mild climate where they can multiply all year long. Mites are very common and can be controlled very easily.

ROUNDWORM

This is the most common of all parasitic worms. It is found most everywhere and needs to be controlled by the bird breeder. The best way to control the Roundworm is to establish a regular system of medication and to keep the pen area clean. We use a regular worming program with our birds which prevent the build up of any of these parasites. Care should be used in determining the dosage as Chinese Painted Quail are very tiny. They need only a fraction of the medication that the much larger birds need. When using a new worming medication, I always try it out on a less valuable bird before giving it to my entire flock.

The life cycle of this pest is such that only the adults can be killed through the use of medication. This makes it necessary to treat the birds again in 10 to 15 days thus ridding the host of the newly hatched young before they reach maturity to spawn more eggs. If this follow-up is not done, the whole procedure is wasted.

TREATMENT: A very successful wormer is put out by several companies is called Piperazine. It is effective and safe up to ten times the recommended dose. Follow the directions carefully.

To assist the birds in their recovery from the stress of worming, it is a good idea to give a high concentrate of Vitamin A for five to seven days after each worming period. This is one of the little things that help.

CAPILLARIA
(Crop worm)

This is by far the most deadly parasite problem. The breeder can be unaware of the problem until his birds begin to die. Several breeders have lost all their birds before the cause was discovered. These breeders were keeping their birds on the ground.

The worms are small and thread-like in nature, and can be found during autopsy in the crop lining and intestine areas.

TREATMENT: To control this worm, the breeder should be very careful because the drugs that must be used are very toxic. There is effective drug treatment available using the product called TRA-MISOL (generic name, Levamisole). Dosage suggested by Pennsylvania State University Veterinary Science Extension is 1 to 2 grams per gallon of drinking water for one day. Repeat monthly for continued control. Set up a "standardized" procedure for keeping your birds free from parasites.

If you keep your Chinese Painted Quail in ground pens beware. In this case, you should treat the soil at least once and year and preferably twice in the following manner. Clean out the top 4 to 5 inches of soil and discard it. Smooth the surface and put down a layer of Sevin dust over the ground. Then put about 1/4 to 1/2 inch of "slacked lime" over the ground. You can buy this from any agricultural supply house. Use only the kind of lime the farmers use. Then put about 4 to 5 inches of sand over the ground. This treatment will last a season and should be repeated every year. Every few days you should rake out the droppings from the loose sand and discard.

CECAL WORMS

This worm is found in the "cecum" or blind gut of the host bird. The adult is about 3/8 to 1/2 inch long and causes little damage to the host. The real damage comes because the Cecal Worm plays a role in the spread of **Blackhead Disease**.

The worm serves as a breeding host for the causative protozoa which reproduces in the ceca and then are passed in the dropping to the ground where they develope into the infective larval stage. Regular worming will prevent a build up of this parasite, especially for poultry and waterfowl.

TREATMENT AND CONTROL: Cecal Worm and Gape worm may be controlled by a number of products on the market. Tramisol is effective in the same dosage as suggested for Capillaria.

TAPEWORM

These are flat, segmented worms that are sometimes found in birds. They are spread by insects and dirty dropping which are ingested by other birds thus infecting the flock.

I have seen some of these worms in nearly all types of birds. In a small Chinese Painted Quail, just one or two of these terrible worms can keep the bird from getting nourishment and the bird will starve to death in a short time. Careful examination of the droppings may reveal segments of this parasite. In this case take immediate action to treat all of your birds.

TREATMENT: A good general wormer can control these worms. Your feed store will have several products that will do the job.

GAPEWORM

This worm lives in the trachea (windpipe) of birds and causes the disease known as the "gapes" due to the gasping of the infected bird. Many infected birds die due to the lack of oxygen. Gapeworms are primarily a problem in young birds raised on the ground. Tilling of the soil helps control this parasite.

HINT: Tramisol is a deadly poison. Do give your birds an overdose as it will kill them.

In all my years with birds, I have only seen this parasite once. It was found in a group of Bantam chicks that came from a very filthy farm. We soon corrected the problem. By the way, the chicks also had Coccidiosis which we had to treat. It seems that many times worm infections are coupled with other disease problems. I suppose the parasite weakens the birds and in turn the infections rage on. It is a safe assumption that your birds will have some type of worm infection. I would suggest that you worm them annually just to be sure. The treatment if administered properly will do no harm at the worst and it could save your birds at the best.

TREATMENT: These worms develope tolerance to a drug in about two years. The breeder should vary his treatment drug to avoid this built up resistance.

Pennsylvania State University Poultry Department suggests using 2 grams of Tramisol per gallon of drinking water for one day, repeat monthly.

HINT: For complete information about the different medications that can be used for worming see pages 252-258 in the authors *Raising Game Birds*. This chart lists all of the available medications and their use.

Appendix

HOW TO GROW SPRAY MILLET

Several years ago we made a delightful discovery, *spray millet that quail just love can easily be grown.* This whole thing came about in our habit of throwing out cage droppings in an open place for the wild birds to pick at. Soon shoots of greenery was everywhere. In just a short time we had spray millet heading out. Before we knew it the heads were getting ripe and we harvested them. An accident got us into a very interesting part of aviculture.

All parrot, canary, and finch breeders know about spray millet. It is the choice feed to get their birds into breeding condition. It can be fed by hanging up the heads to the cage, soaking it and letting the birds munch on it, or just thrown into the bottom of the cage. Chinese Painted Quail just love it as a treat food. It also helps get them into breeding condition.

If you want to have some fun, try growing some spray millet. It is really very easy to do.

"The first requirement for a successful crop is the proper land. Spray millet does best in well drained soil with much organic matter" (Underwood, 1982).

Hard packed soil with standing water is least desirable. The seeds will have a hard time sprouting in the hard soil and the standing water will cause them to rot. If your soil has lots of clay you need to get some kind of mulch material mixed into it. Peat works well and can be bought rather cheaply. The more organic matter in the soil the better. The larger heads are produced in the loose, organic rich soils.

HINT: Prepare the soil in late fall and let it stand all winter. Put some kind of mulch on the surface to help hold in the winter moisture.

After the land is ready it is time to think about the seeds. You can get spray millet in nearly any pet or feed store. You can pick out the heads that are the large and filled with seeds. Usually, you buy spray millet by the head or by pound weight. Choose heads that are rich brown in color. Be sure that the seeds are still in the head and have not fallen out because of being overripe. The ideal color is to let them get as brown as possible before falling out of the seed heads. You will soon learn when to harvest.

If you have never grown spray millet before, you need to give some thought to the spacing of the plants in your field. The rows can be about two feed apart which gives room for the plants to grow and at the same time offers group support to the plant in high winds. The spacing of the plants in the rows is determined by the expected harvest. If you want lots of smaller heads then put the plants about two inches apart. If you want the larger heads then put the plants about six inches apart.

Plant the seeds in nice even rows in shallow trenches. When they sprout they can be thinned to the proper space. Normally, all of the care necessary is watering, weeding, and heeling up dirt around the base of the plants to keep them from falling over. Some of the plants will grow taller than others. Some will produce more seed than others.

Plant some spray millet seeds about every three to four weeks during the growing season. This way you will always have seed heads in the milk stage.

Regardless of the amount of time required to grow millet, I feel it is worth it!

HOW TO
SPROUT
SEEDS

Every keeper of birds knows the value of sprouting seeds in the diet. It has been proven through much research that sprouted seeds are beneficial to all birds. Breeders of parrots especially know the value of sprouted seeds in getting their birds ready for the breeding season.

Sprouting seeds is very easy and can be great for Chinese Painted Quail. At first, they might not take to this new kind of feed, but soon will be waiting for this highly nutritious food.

HINT: Use only fresh seeds for sprouting. Stale or old seeds do not sprout.

For a long while we just could not get our seeds to sprout before they "soured." We used the mold inhibiter called Calcium Propionate and it did not help. I was about to give up. Then it occurred to me that perhaps the seeds that we were using were not fresh. We started over with more seeds that were from a different source. That did it! We have had no trouble since as we always use fresh seeds that germinate without getting sour. I have since read that the people that raise parrots use the sprouting technique to test their seeds for freshness. If they do not sprout they have lost much of their food value and should not be used.

The kind of seeds you sprout is up to you. We use good fresh canary and finch mix, safflower, even sunflower seeds. Stay clear of the bargain bags of "wild bird seed" that you can buy very cheap. Get the best available mix to give your birds an extra bit of nourishment.

Here is how we do it. First, we gather together some items for the procedure. We use three or four of the plastic refrigerator bowls that can be sealed with an air-tight lid. We like the size made to hold sandwiches for storage. They measure about 5 X 5 X 1 1/2 inches. You can use whatever size you want. The reason we use three or four of these bowls is we start a new batch every week which gives us a continuous supply of sprouts.

Place enough seeds in one of the plastic containers to about 1/3 full. Soak these seed for 18 to 24 hours. It is good to give them a good wash in running water for about three minutes every eight hours or so. Rinse them until the water runs clear.

HINT: In cold weather it may take three or four days for the seeds to sprout. Not as long in warmer weather.

After the soaking period, drain them well and place them in a warm place. We put the lid on loosely so the moisture is held in. Do not make them air-tight as they will sour for sure. Let the air to circulate in the container. Put in a warm, dark place and in a day or two you will see the seeds put out tiny roots. Loosen the seed very gently and if necessary put some more water in the container to replace the moisture that has evaporated. You can feed the sprouts to the birds when the tops are green and the shoots are tender. They may smell a little sour until they begin sprouting and then all of the sour odor disappears. If you wash them well every eight hours during the soaking period you should have no problem assuming you have fresh seeds.

I am totally convinced that feeding the variety of foods to your birds will pay dividends over and over. The extra spark of health, the brightness in the eyes, the sheen of the feathers all comes from giving your birds the little things that mean so much!

Sources

Allen, George A. Jr. "Ornamental Pheasants." *The Game Bird Breeders, Aviculturists, and Conservationists' Gazette*, Vol. XVI, Jan.-Feb. 1967: 34-41.

Black, Robert G. "Finches in the Outdoor Aviary." *American Cage-Bird Magazine* December 1984: 22.

Flemming, Robert L. Sr., Robert L. Flemming, Jr., and Lain Singh. Bangdel. 1979. *Birds of Nepal*, Kathmandu: Avaloc Publishers, 70.

Johnsgard, Paul A. 1988, *The Quails, Partridges, and Francolins of the World*, Oxford; New York Tokio:Oxford University Press, 35-37, 202-205.

La Rosa, Don 1973, *How To Build Everything You Need For Your Birds*, Simi: La Rosa Publications, 23-32.

Rutgers, A. 1973. *The Handbook of Foreign Birds in Colour*, Vol. 2, London: Blandford Press, 123.

Schroeder, Dick. "Lively, Lovely Lories." *Bird Talk* March 1989: 44-52.

Smith, Dr. Tom. "Quail Quill." Letter of Mississippi Cooperative Extension Service, No. 42, May, 1989.

Underwood, Michael, "Growing Spray Millet." *American Cage-Bird Magazine* August 1982: 24.

Wilson, W. O. "Cannibalism in Game Birds." *Game Bird Breeders, Aviculturists, Zoologists and Conservationists' Gazette* February 1972: 15-16.

Woodard, A.E. "Cannibalism: Its Causes and Control." *Game Bird Breeders, Aviculturists, Zoologists and Conservationists' Gazette* April 1977: 9-10

Advertisements

I have dealt personally with each of these companies and have been treated fairly in every instance. I therefore, can recommend each of them based on my personal experience. I list them alphabetically.

ACADIANA AVIARIES is owned and operated by Garrie Landry. This man is setting the pace in the US by his extensive breeding and obtaining new mutations of Chinese Painted Quail. Garrie Landry has a very informative price list sent on request. Be sure and send a SASE.

BROWER MANUFACTURING is one of the leading manufacturers of poultry equipment. Many of their products are helpful and necessary in the raising of game birds. They will be glad to supply you with a catalog of their products.

GAME BIRD BREEDERS' GAZETTE is the worlds leading game bird magazine. It has been in publication for nearly 40 years. It includes articles on every aspect of game bird propagation has the best color photographs found in any publication.

GAZETTE INDEX is a listing of every article printed in the magazine. One of the most helpful services offered is a **REPRINT SERVICE** where anyone can order a reprint of any of the 2197 articles. Cost of each reprint is only $1.50 each.

GAZETTE **MEDICATIONS DEPARTMENT** offers hard-to-find game bird medications. They offer same day mailing service which is very important when the medication is needed in a hurry.

LYON ELECTRIC, INC. has been offering the game bird breeder products for many years. They offer a complete source of products needed for successful game bird propagation. They send a free catalog on request.

RAISING GAME BIRDS has become a very popular book. It will be translated into Spanish, Russian and German and is sold all over the world. The information in this book can be used for any type of game bird.

RAMONA BIRD FARM offers a complete selection of game bird products along with selections for the cage bird hobby. They also offer for sale many species of birds.

SNYDER SUPPLY (Bookseller) offers a complete source of every book that has ever been written about birds. If they do not have it -- they can get it for you. This source for bird books is most welcome.

Acadiana Aviaries

Garrie P. Landry
Rt. 1, Box 199   Chatsworth Road
Franklin, Louisiana   70538

Specializing in all varieties of Button Quail

*Normal Wild Type, Silver, White, Fawn,
Blue Faced, Blue Faced Cinnamon
Pearl, Silver Pearl, Cinamon Pearl
All Red-Breasted Varieties
Plus More!*

Mature stock and some eggs available year round!
Shipped Next Day Service, Nationwide at Lowest Cost!

ACADIANA AVIARIES is one of the South's Largest
Bird Farms

Also Specializing in:
- ❑ Largest U.S. selection of Zebra Finch varieties**
- ❑ Largest U.S. selection of Diamond Dove varieties**
- ❑ Many rare and exotic species of Doves and Finches
- ❑ Roul Roul Partridges
- ❑ Rare Central and South American Currasows

**Many of these from imported show winning stock!

Visitors Welcome by Special Appointment

"Call or Write for Descriptive Price List and Shipping Information
(When writing please send 2 stamps for prompt reply)

Garrie Landry, Owner
Chris Rogers, Curator

Phone (318) 828-5957

COMPLETE QUAIL EQUIPMENT NEEDS

INCUBATORS

845 Incubator

By **Brower**

RT20 Plastic
Reel Top Feeder

FEEDERS

209 Round
Galvanized Feeder

MJ9 Round Galvanized
Fruit Jar Feeder

FT220 Plastic
Flip Top Feeder

WATERERS

0 Galvanized
Fruit Jar Fount

QB555 Drown Proof Quail Base
with 45J 1 Gallon Jar

75 Drown Proof Base
for 1 Quart Jar

1QW Plastic
Super Start Waterer

BROODERS

UA4T Four Bulb Brooder

6402 Box Brooder

UA1 Single Bulb Brooder

Brower • P.O. Box 2000 • Houghton, Iowa 52631 USA • 319-469-4141

150

If you are interested in quail, partridge, pheasants, waterfowl or other game birds . . .

Subscribe to the
Game Bird Breeders
GAZETTE

It is the oldest and largest game bird magazine in the world. Every issue contains interesting articles on how to keep and raise all the different kinds of wild waterfowl, pheasants, quail, partridge, grouse and other varieties of game and ornamental birds. Highly illustrated with dozens of photographs and drawings each month, some in full color, showing what all the different species look like. It also has a wildlife supermarket where thousands of all kinds of game birds and animals are offered for sale, trade or wanted. You can sell or buy anything you want through the GAZETTE. The best and most beautifully illustrated magazine for both commercial and ornamental breeders. Instructive, informative. Invaluable to both beginning and veteran breeders alike. Now in its 38th year of publication.

Subscription Order Blank
Send this blank with remittance to:
The GAZETTE
1155 East 4780 South, Salt Lake City,. Utah 84117
Subscription Rates:
United States & Canada: $20.00 per year, $37.50 for two years.
Here is my check for ☐ *one year* ☐ *two years subscription to the GAZETTE.*

My name is: ....

Address ....

Be sure to include zip code **151**

THE GAZETTE INDEX

Compiled by
Leland and Melba Hayes.

Contains listings of over 3,000 articles from past issues of the *Gazette*

Almost 50 pages of information. Complete ordering information is sent with your order. Find articles in your back issues of the *Gazette* easy.

Order reprints of that article about a new species.

ORDER YOUR COPY TODAY !

$12.00 POSTPAID

Send check to:

Dr. Leland Hayes
Dept. BQ
P.O. Box 1682
Valley Center, CA   92082

VIONATE

The Miracle Supplement

For raising Pheasants, Quail, Partridge, Waterfowl, Cranes and other Game Birds.

Vionate is the best and leading viatmin-mineral supplement available for birds. It helps keep birds in peak condition and is especially effective against preventing stress, disease, crooked toes, leg problems, etc. Comes in powder form — mixes easily in feed. Contains 21 essential vitamins and minerals. For baby, growing, and adult birds of all kinds. Improve your hatches by feeding to breeding stock and increase liveability and growth of chicks by adding this to their diet.

8 ounce can . . . . **($7.25 plus $1.50 postage) $ 8.75**

32 ounce can . . . .**($19.00 plus $4.00 postage) 23.00**

Orders filled and mailed out same day received.

Order from: GAZETTE Medications Dept.
1155 East 4780 South, Salt Lake City, Utah 84117

153

Marsh Farms
INCUBATORS & BIRD EQUIPMENT

JUST FOR YOU BUTTON QUAIL BREEDERS!

We have designed a new Roll-X2 incubator with specialized button quail grid.

•This effective incubator has 305 egg capacity and consists of a full automatic turner, temperature and humidity control.

TURN-X

WE ASLO CARRY THE FINEST:
- **BROODERS**
- **FEEDERS & FOUNTAINS**
- **EGG CANDLERS**
- **HOW TO BOOKS & MORE**

Ideal for the
Serious
Hobbyist
Specialized
Breeder or
School
Program

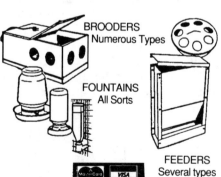

BROODERS
Numerous Types

FOUNTAINS
All Sorts

FEEDERS
Several types

Write for our
CATALOG today.

MANUFACTURED BY

LYON
ELECTRIC COMPANY, INC.

2765 MAIN STREET
CHULA VISTA, CA 92011 U.S.A.
TELEPHONE: (619) 585-9900

RAISING GAME BIRDS

By
Leland and Melba Hayes

IF YOU LIKE QUAIL, PHEASANTS, PARTRIDGE . .
.

. . . ORDER YOUR COPY OF THIS BOOK THAT IS VALUABLE RESOURCE READING !

300 Pages of information and photos for the beginner as well as seasoned aviculturist. Contains major sections on Species of quail, partridge and pheasants, commercial game bird breeding and health and diseases of game birds.

GET YOURS TODAY!

LEARN MORE ABOUT THESE BIRDS!

Send $20.00 to: Dr. Leland Hayes, Dept. BQ
P.O. Box 1682
Valley Center, CA   82082

RAMONA BIRD FARM
1798 Keyes Road
RAMONA, CA 92065
(619) 789-2473

Established in 1976

WE ARE BREEDERS!

HAND-FED BABIES!

Birds from...
Africa, Indonesia,
South America

GAMEBIRDS
WATERFOWL
PEACOCKS
QUAIL

Cages, Supplies, Books, Feed

WEEKEND PETTING PLAYPEN WITH THE PARROTS AND EMUS

OPEN 7 DAYS A WEEK, 9 AM TO 5 PM

"Follow the signs from downtown Ramona..."
(Start at Main and Hwy 78)

SNYDER SUPPLY
(Bookseller)

The most complete line of New and Used Books on:

Game Birds-
    Waterfowl-
        Poultry-
            Pigeons-
                Cage Birds-

Also complete inventory of Books on:

Mammals-
    Reptiles-
        Wildlife-

For information contact:

        DeWayne Snyder
        426 Beech
        Ottawa, Kansas 66067

        (913) 242-1805

Glossary

Assembly Line Method -System used by author to raise chicks.

Bacteria -Microscopic single-celled forms.

Beetle -Adult stage of the mealworm.

Brooder -Environment in which chicks are raised.

Cannibalism -The picking (pecking) of birds by other birds.

Carrier -An apparently healthy bird that harbors disease organisms and is able to transmit to other susceptible birds.

Ceca -Blind pouches at the junction of the small and large intestine.

Coccidiostat -Any group of chemical agents mixed in feed or water to control coccidiosis.

Contagious -Refers to infections that may be transmitted from one individual to another.

Culture (Noun) -Medium in which mealworms grow.

Debeaker -Machine used to trip beaks of birds.

Forced air -Type of incubator in which air is circulated.

Fumigation -The killing of germs in an incubator or brooder using chemicals.

Hardware cloth -Welded wire mesh, comes in various sizes.

Hemipode -Family of birds closely related to cranes.

Hygrometer -Instrument used to measure humidity in an incubator.

Incubator -Environment which keeps correct conditions for embryos to develop.

New World Species -Quail species native to North and South American continent.

Nominate race -Prominent head of family which begins the classification chain.

Parasite -Animal form that lives on or in a bird.

Primary wing flights -The seven or so feathers at the end of the bird's wing.

Symptom -Detectable signs of disease.

Wet Bulb -Term used for the hygrometer which is used to measure humidity in incubators.

Index